EVADING AND ESCAPING CAPTURE

URBAN ESCAPE AND EVASION TECHNIQUES
FOR CIVILIANS

SAM FURY

Illustrated by
NEIL GERMIO

NONFICTION
S F
BOOKS

WARNINGS AND DISCLAIMERS

The information in this publication is made public for reference only.

Neither the author, publisher, nor anyone else involved in the production of this publication is responsible for how the reader uses the information or the result of his/her actions.

CONTENTS

EVASION PLANS

ESCAPING CAPTURE

PRELIMINARIES

CARS

NEGOTIATION

THANKS FOR YOUR PURCHASE

Did you know you can get FREE chapters of any SF Nonfiction Book you want?

https://offers.SFNonfictionBooks.com/Free-Chapters

You will also be among the first to know of FREE review copies, discount offers, bonus content, and more.

Go to:

https://offers.SFNonfictionBooks.com/Free-Chapters

Thanks again for your support.

INTRODUCTION

In this book, you'll learn the skills you need to prevent and escape capture. It's packed with covert military and spy escape techniques adapted for the average civilian.

Anyone has the potential to be taken, though some people are more of a target than others.

- **Women.** The primary targets for sexual predators, and the most likely hostages in crimes gone wrong.
- **Children.** Targets for sexual predators and/or likely to be held for ransom.
- **High-profile individuals** (politicians, celebrities). Held for ransom.
- **Tourists.** Western tourists are more at risk for being held for ransom in developing and/or politically unstable countries.

Applying what you learn in this book will keep you and your loved ones safe, whether at home or abroad. It also contains information for preventing common theft and other crimes.

This is a two-part book.

Part 1. Evading Capture

This part will give you the tools you need to avoid becoming a victim.

There are five key rules for preventing capture:

1. Be aware.
2. Avoid danger.
3. Remove temptation.
4. Plan and prepare.
5. Keep handy things nearby.

Rules 1, 2, and 3 prevent things from happening. Rules 4 and 5 ensure you're ready in case something does happen.

Part 2. Escaping Capture

When the five rules for evading capture fail, use the information in this part to plan and execute your escape.

It includes specific escape techniques (breaching locks, improvising explosives, stealth movement, hostile negotiation, etc.) and other relevant information.

Action and Adaptability

Action and adaptability are how you use what you learn from this book.

When it comes to evading and escaping capture, the sooner you act, the better. This is true at every stage of the process:

- Act on learning about and training yourself in evading and escaping capture.
- Act on using what you learn in part 1 to keep you out of danger.
- Act fast when you are in danger (using the skills in part 2) for the best chance of escape.

The concept of acting fast when in danger is important. Delayed action results in a loss of opportunity. As soon as you spot the warning signs, distance yourself. Stay calm and follow your plan.

When taken, escape early. The more time that passes, the harder it becomes. Security will increase, tools will be confiscated, and your strength (mental and physical) will deteriorate.

Adaptability is the ability to apply what you learn to specific situations. Things will never go exactly as you plan. Be ready to overcome any obstacle that presents itself.

EVADING CAPTURE

BE AWARE

Being aware is about actively noticing what is happening around you.

This has two major benefits:

- It allows you to spot early warning signs of potential dangers.
- It makes you a less appealing target.

CULTIVATING AWARENESS

Cultivating awareness isn't hard, but it takes discipline to not be distracted.

One way to do it is to talk to yourself silently. As you look around, tell yourself what you see, hear, smell, etc. Things that are useful to take note of include:

- Agitated animals. Animals often sense danger before humans do.
- Landmarks. Take note of landmarks for orientation and to use as rally points.
- Exit routes. Always know your best route(s) of escape.
- Potential hazards.
- Potential weapons. Go for them if you are attacked.
- Suspicious people. Look out for nervous or "shifty" behavior.
- Suspicious vehicles. Memorize license plates and general descriptions.
- Strange sounds, smells, etc.
- Unusual traffic. This may signify people running away from danger.
- Anything else out of the ordinary.

Always place yourself where you can best observe your surroundings, such as with your back to the wall and facing the entrance.

When you need to focus your attention, such as when you're on your phone or during a conversation, look around every 10 seconds to ensure you're safe.

Increase other people's awareness of you. Let trusted people know your itinerary, and check in with them. Ensure they know what to do if you do not check in.

Cultivating a constant awareness of your surroundings and yourself can be difficult at the start. It takes a lot more brain power than daydreaming or staring at your phone, but with practice, it will become second nature.

Related Chapters:

- Rally Points

AVOID DANGER

Criminals can be anywhere at any time, but there are times and places where your risk of danger increases. Here are some examples:

- Daytime is generally safer than night time, due to lighting.
- Crime rates go up during the holiday season, and elections can cause social upheaval.
- Isolation makes you an easy target, but areas that are too crowded make it easier for petty thieves to rob you.
- A cubicle in the men's bathroom is safer than the urinal.
- Geographical location (Canada vs Somalia, suburbs vs slums).
- Checkpoints/roadblocks.
- ATMs on the street. Go inside a mall or bank instead.
- Alley entrances and other hidden areas.

When you have the choice, go for the safer option. Here are some general guidelines for doing that:

- Avoid isolated areas, even those inside buildings, such as laundry rooms, mailrooms, or parking garages.

- Research public transport schedules to minimize your wait time on the street.
- Sit near exits while being able to see who comes in.
- Choose aisle seats.
- Stay near "safe" people, like, security guards, family members, or bus drivers.
- Keep to well-lit areas.
- Avoid potentially hostile crowds, such as those made up of drunk people or male youths.
- Walk facing traffic.
- Stick to the center of the sidewalk, so you're not too close to passing cars or ambush spots.
- The second and third floors of buildings are the safest in, especially in hotels and apartment complexes. The first floor is not secure, while if you're on the fourth floor or higher, fire ladders may not reach you.
- Rooms near fire exits and elevators are good, but those near stairwells are not.

Whenever you feel in danger, get to a safe spot. A safe spot is anywhere that has people, cameras, and/or good lighting. The more of these things, the better. For example, find a:

- Police station.
- Supermarket.
- Shopping mall.
- Gas station.
- Busy restaurant/cafe/bar.

Take the time to locate the all-night safe spots in your area.

Related Chapters:

- Elevators

TRAVELING

Avoiding danger while traveling takes a little extra work. The main thing is to gather knowledge. Before you go, research your destination, the local customs, and local scams. Avoid dangerous areas and do as the locals do where possible. Eat how they eat, for instance. Subscribe to travel warnings, which you can do here:

https://subscription.smartraveller.gov.au/subscribe

Once you're at your destination, making friends with a trustworthy local is a good way to gain insider information. They can identify local hazards, recommend places, tell you how much things should cost, etc.

Be careful of who you make friends with, however. Local customer service staff (hotel receptionists or waitresses at local coffee shops, for example) are usually safe, but you can never be sure. Do not give them the details of your timetable or other sensitive information.

When you're interacting with the locals, using a few words in their language with a genuine smile can get you a long way. Learn how to say "Hello," "How much does it cost?," "Thank you," and "Good-bye." Avoid conversation about religion, politics, and money. If they bring one of these subjects up, be respectful.

When in high-risk areas, stay away from places that are frequented by foreigners (hotels, attractions, restaurants, markets, etc.). Opt for the "non-foreign" equivalents instead. As a bonus, the local options are usually better, being less crowded, cheaper, and more authentic.

If there is a terrorist attack, stay away from the embassies for a few days in case of a follow-up.

Related Chapters:

- Common Scams and Petty Theft

DIGITAL SECURITY

These days, any amateur hacker can steal your information with basic software. Use the following tips to deter online stalkers and scammers.

Computer/Laptop

Cover your camera with non-transparent tape in case someone remotely accesses it or you forget to hang up after a video call.

Update your software whenever there's a new patch.

Log off your online accounts (banking, shopping cart, etc.) and your computer when leaving it, especially in public places such as at work or school.

Disable unused USB ports to prevent hackers from using "plug and play" hacking gadgets such as keyloggers, bash bunnies, and rubber duckies.

WIFI

Never use a free, "no login" open hotspot. It may be one that a hacker set up using a WIFI Pineapple or some other device.

Use a VPN to encrypt your activity on any network that isn't your personal one, and especially when banking, shopping online, or sending sensitive information over email. For an extra layer of security, use the TOR browser:

https://www.torproject.org/download

Stay off the dark web, even when using a DSL internet connection.

Password Security

A good password is secure, unique, and easy to remember. Here is a method for making multiple secure passwords.

Choose a random word that is at least 6 characters long, such as "Panasonic."

Change the word into a mixture of upper and lowercase letters, replacement letters, symbols, and numbers. Have at least one of each. In this example, you might end up with "P@nas0n1K."

This is your base password.

Add a prefix or suffix to your base password for each account. Use a specific prefix/suffix that relates to each account, but use the same pattern for all accounts.

For example, using the base password above and the following companies, take the first and last letter of each company name as a prefix for its specific password.

- Merrel - MLP@nas0n1K
- Chase - CEP@nas0n1K
- Robinhood - RDP@nas0n1K

To make it even more secure, use more letters or add another level. For example, you could use the number of letters in the company name as a suffix at the end, as shown below:

- Merrel - MLP@nas0n1K6
- Chase - CEP@nas0n1K5
- Robinhood - RDP@nas0n1K9

If you think you'll have trouble remembering it, you can record the base word somewhere in its original form (e.g., "Panasonic"). Make sure you put it somewhere safe—not in your wallet or near your computer. Now you only have to remember the pattern.

You should also change your passwords at least monthly. Streamline this by doing the following:

- Make a list of all the places you need to input your password.
- Choose one day a month when you'll go through and change them all.
- Change your base password and the prefix or suffix pattern.

If you think your base password and/or pattern has been compromised, change all your passwords ASAP.

Never tell anyone your passwords!

Besides making a secure password and changing it regularly, there are a few other things you can do to increase your login security.

Take advantage of one-time passwords (OTPs) sent to your phone or via a third-party gadget or app, fingerprint recognition on your phone attached to your account, and whatever else is offered.

Security questions are a basic layer of added protection. Make them even more useful by lying in response to them. You can use an opposite answer, a variation of your base password, or a jumble of the actual answer.

Social Media

The safest thing is to not have any social media accounts, but that isn't practical for many people.

The next best thing is to be mindful of what you share and who you share it with.

- Be selective with who you become "friends" with.
- Don't post any personal information.
- Don't post anything that gives away your location, routine, or when you are absent from your home.

Email

Encrypt any sensitive information.

Delete emails from untrusted sources without opening them. Signs of scam emails include provocative subject lines, nothing but a link in the body of the email, emojis in the subject line, and jumbled letters as the "From" address.

Do not open or download any suspicious links or attachments.

Beware of impersonations, such as emails from "your bank." Never give personal information by replying to an email or log in to your account via a link in an email. Instead, contact the relevant company independently (e.g., by phone or via its website).

Online Shopping

When you're shopping online, check out as a guest. If you register on a company site and it's hacked, your information will be compromised. If you want the "free gift" for signing up with a company, use a temporary email from a service like Guerrilla Mail:

https://www.guerrillamail.com

Phone

Get an unlisted number.

Answer with a simple "Hello" instead of your name and/or number, and do the same with your voicemail.

Don't give your personal details to anyone you don't know who calls you. If they say they're from a company, ask for a reference number and call the company back. Find its number yourself via its official website.

Hang up on any menacing calls. Don't react. If they persist, call the police.

To truly prevent tracking (by the government, for example), take the SIM card and battery out of a burner phone (if possible) and smash the phone after you've used it.

Update your phone software.

Use a VPN.

Never tell a stranger who calls you that you're alone.

Related Chapters:

- Stalkers
- Tracking

COMMON SCAMS AND PETTY THEFT

The chapter highlights common methods criminals use to abduct, rob, and/or scam people, and what prevention methods you can use.

It is not intended to make you cynical about everyone. Most people are not "out to get you," but it's wise to not be too trusting, either. Use common sense depending on the situation you are in.

Distraction

A distraction is engineered so your attention is focused elsewhere—on someone causing a scene, for example. While you're distracted, your property is stolen or you're attacked. Situational awareness will help you combat this.

Collision

Someone will purposely bump into you, but make it seem like it's your fault. In the process, he will drop or break something "of value" and demand that you pay for it.

A more elaborate version of this is someone running out in front of your car.

This can also happen in a car where the scammer will bump your car with theirs. When you get out, a third person will steal your car or you will get abducted.

If this happens to you, tell the person that you were being very aware and that it was not your fault. Be firm and polite. If he persists, call the authorities.

Sometimes the cost may not be worth the hassle. If he wants a relatively small amount, consider paying.

When in a car, don't get out. Put on your hazard lights and call the police. Note down the other car's plate number, driver description, etc. If someone approaches your car, tell him to follow you to a safe place, such as a police station or populated area.

Honey Pot

In this scam, an attractive person (the honey pot)will befriend you. After some time, you will go to a restaurant, bar, or something similar, where there will be no prices on the menu. At the end, there will be a hefty bill, and the "honey pot" will get a cut.

To prevent this happening to you, when you want the "local experience," take your advice from a guidebook, not someone you just met. Also, never order something without knowing the price.

Impersonation

In this scenario, someone dresses up in an "official" uniform and tries to gain access to your house or get your personal information. This may also happen over the phone or online, when someone pretends to be a co-worker, government official, bank worker, etc.

Your best defense against impersonators is to trust your instincts if something doesn't seem right, and to beware of anything that's too good to be true.

Never give your personal information to someone who has called you. Instead, hang up and call the business they claim to represent. Verify people are who they say they are. For example, if the gas man is at your door, call the gas company to confirm that.

The Good Samaritan

When you see someone in distress, it's natural to want to help them, but be careful. Often the "damsel in distress" routine is a setup, and you will get pickpocketed, mugged, car-jacked, or worse.

Always be extra careful of helping someone when you're in an isolated area. If it's a car issue, the best thing to do is call emergency services.

If you do decide to help someone, be aware of your belongings and your surroundings in case an accomplice attacks you.

Don't hesitate to leave if you sense something isn't right. Verifying the situation by asking questions about the victims' story will help you decide whether it's a scam or not.

Reverse Good Samaritan

Here, the scammer plays the role of good Samaritan. For example, they might try signaling you to pull over because something is wrong with your car. Unless there's an obvious emergency, wait until you're somewhere safe to check out the issue.

The Box

In this situation, several criminals surround you. This can be while you're walking or in your car. To prevent this from happening to you, always leave room for escape and ram your way out if necessary.

Taxi Scams

There are many types of taxi scams. They can also apply to any private form of transportation, such as tuk-tuks. Some common examples are:

- Taking you on the longest route possible.
- Putting the meter on "suburban rate" rather than the standard rate.
- Taking off with your luggage in the trunk when you get out of the car.

- Taking you on a detour where their criminal friends are waiting.

There are several things you can do to protect yourself from taxi scams:

- Use official taxis or ride services such as Uber or Grab.
- Opt for public transport. It is often safer and always cheaper.
- Don't use touts. Select your own taxi or use the official taxi line.
- Travel light so you don't have to put anything in the trunk.
- Know the route/track yourself with GPS.
- Only go to your original destination. Don't let the driver take you to a "better" or "cheaper" place.
- Ensure there is a handle on the inside of the door before getting into a car.
- Use the meter.
- Check the driver ID matches the driver.
- Do not share taxis.
- Keep the windows up and doors locked.
- Ask another person, such as your hotel concierge, how much the trip would normally cost.
- If you have a small fare dispute, it is often better to just pay it.
- Note the license plate number (or take a photo) and send it to a trusted friend or family member so he/she can track your movements. Take a photo of the driver as well. Let him see you do this. You can ask if it is okay to do so first. If he objects, call a different service.
- If a taxi driver refuses your directions, get out as soon as he stops in traffic.

Pickpockets

A pickpocket is a skilled thief who steals things out of your pocket or bag.

Here are some tips for combating pickpockets:

- Your front pants pockets are the safest place to keep things.
- Avoid loose pockets.
- Use zipped or button-up pockets.
- Putting an elastic band around your wallet will make it stick to your pocket.
- Don't leave anything unattended, especially on the beach.
- Be cautious in crowds, at ATMs, and when distractions are created.
- Don't check your wallet constantly, as this is a tell-tale sign of where it is.
- Hide the bulk of your cash in a money pouch around your neck or in a secret pocket in your pants. Keep enough money in your pocket so that you don't have to reveal your secret spot in public.
- Only access your secret spot in private (inside a bathroom cubicle, for example).
- Watch your watch, especially during handshakes.
- Confronting a pickpocket (not a mugger) in public is generally safe. He will deny any wrongdoing, but will probably not be violent. Get a good description of him for the police.
- If your wallet gets stolen and you get a call from the police to collect it, always call them back to verify they have it. The thief may be getting you out of the house. If your wallet gets "found" by a stranger, you should still cancel your credit cards.

Bag Snatching

A bag-snatcher is more blatant than a pickpocket, and more danger-ous. He is likely to fight to escape since he can't deny the crime.

Bag-snatching is a generic term. It applies to anything you carry such as a purse or phone.

To protect yourself from a bag snatcher or someone pinching some-thing out of your bag:

- Use straps and sling it across your body so it is at your front.
- Hold it close and tight.
- When you're walking, carry it on the side away from the street.
- Ensure it's closed properly.
- In a bathroom stall, keep it away from the door and the open space under the stall. Choose a stall with a solid wall on one side.

Conmen

A conman/woman is someone who gains your trust and then takes advantage of you. After building rapport with you, a conman may use one or more of the following psychological tricks.

Reciprocity. When someone gives you something, you're more likely to give something back in return. This may be a favor, a gift, money, information, etc. The conman will give you something and want something bigger in return.

A variation of this is for him to help you without you asking and then expect money. This is common when traveling. For example, you may encounter unofficial porters.

Small Request. The conman may start making small requests that you are likely to grant. As you get used to giving, the requests will increase in size. A variation of this to ask you for something big.

When you refuse, he'll ask for something more reasonable, which is what he really wants.

Bandwagon. People naturally want to do things that other people are doing. The conman will imply that "everyone else is doing it" and you should too.

Scarcity. This scam plays on your fear of missing out—the feeling you had better do or buy something soon, before it's no longer available.

Related Chapters:

- Traveling
- Picking Pockets
- Detect Lies

SAFE ROOMS

A safe room is a fortified place inside your home where you can take refuge in the case of an intruder or disaster situation.

You don't need a purpose-built safe room. Here's how to make one.

Choose the Room

Use any room that is accessible to all household members. Consider people with limited mobility, such as the elderly, the disabled, and children. The room must be able to lock from the inside, but stay unlocked so everyone can access it in an emergency.

A room with few entry/exit points is best.

Secure the Room

Make the following modifications so you can secure the room from the inside:

- Solid door.
- Deadbolt.
- Extra barricades on doors and windows.
- Something to hide behind if shots are fired.

Stock the Room

Have enough supplies to last your family at least three days, as well as some security and rescue items. At the very least, include the following:

- Cell phone and charger.
- Flashlights and spare batteries.
- First-aid kit and prescription medication.
- Non-perishable food.

- Water for drinking and hygiene.
- Sanitary products.
- Buckets and garbage bags for ablutions.
- Weapons. (Store them appropriately.)
- Security camera feeds.

Related Chapters:

- Secure Entry Points

DEALING WITH POLICE

Unless you're a victim in need of emergency help, it's best to stay away from the police. There's a fine line between "witness" and "suspect," and once they decide to capture you, your chances of escape are slim.

The number-one rule for dealing with police is to not volunteer any information unless it will lead to the immediate capture of a dangerous criminal (the direction in which a shooter ran, for example).

Be extra careful of police and other government services in times of civil unrest. It becomes an "us and them" culture, and they are a trained group with weapons.

Here are some general do's and don'ts for dealing with hostile police. Take into consideration the specific country and situation you are in.

Do:

- Keep your hands in view.
- Ask if you are being detained. If not, walk away. If you are, stay in one spot until instructed to move.
- Give them your ID if asked.
- Know your civil rights in the country you are in (those relating to search and detainment, for instance).
- If the police are at your home with an arrest warrant, go out and lock the door behind you.
- Notify people of your arrest/detainment—the more, the better.
- Ensure everyone involved knows to keep silent.
- Record your interactions with police in writing and/or on video.

Do Not:

- Run or resist arrest, except in special circumstances.
- Touch police officers or their equipment.
- Make sudden movements.
- Be rude. Instead, politely say, "Sorry, I have nothing further to say."
- Agree to go to the station, unless they are arresting you.
- Inject yourself into any situation unnecessarily.
- Consent to a search of your person, home, car, or office. If they conduct one, say forcefully, "I do not consent to this search," but do not resist physically.
- Confess to anyone. Other inmates may be informants. Never discuss your case with anyone but your lawyer.
- Fall for their interrogation tactics.

Common interrogation tactics include:

- Lengthy detainment.
- The claim that they have evidence, so you may as well confess.
- False charges for not answering questions.
- The claim that friends have cooperated or turned on you.
- Lighter punishment for a confession.
- A "good cop, bad cop" routine.

Traffic Stops

When a cop signals for you to pull over:

- Turn on your hazard lights and drive slowly to a safe place. A safe place is one away from traffic, where it is well lit and there are witnesses.
- Stay in your vehicle, turn off the radio, turn on the interior light, and place your hands on the steering wheel.

- Only move when asked to, and move slowly.
- Never admit to an offense. If asked if you know why you're being pulled over, say no.
- Do not contest any citation the officer gives you. Take it to court later instead.

DISAPPEAR PERMANENTLY

There are a few reasons why you may want to disappear permanently You may need to hide from the government, gangsters, or a stalker, for example. If you're thinking about doing so, here are some things to consider.

Where to Go

Who you're hiding from will determine how far you have to run—to a different city, state, or country. Choose somewhere unexpected so no one will think to look for you there.

When fleeing from the law, go somewhere that has no extradition treaty with your country, and where there's less government control. Southeast Asia or Latin America might be good options. Time is crucial; you have to leave before you're put on any no-fly lists.

Cut Social Ties

It's preferable to do this slowly, so when you finally do disappear, people won't raise the alarm.

- Start seeing friends and family less and less, until not hearing from you is considered normal.
- Delete your social media accounts.
- Quit your job officially, so no one will get concerned when you don't show up.
- Tell anyone who might worry that you're going on an extended vacation and won't contact them. Use the excuse of an electronic detox.

Traveling

Make dummy travel plans with credit cards, then carry out your real plans with cash.

Hiding Your Identity

Once you officially "disappear," you need to keep your old identity secret.

- Withdraw all your money in stages before you leave, and pay cash for everything.
- Burn all your ID, bank cards, etc.
- Stay out of trouble.
- Never go anywhere or do anything where someone might ask you for ID (no driving).
- Rent directly from places with "for rent" signs and be a perfect tenant.
- Avoid security cameras. If you can't, keep your head down and wear sunglasses and a hat or hoodie.
- Run water if you suspect listening devices.

Contacting Home

Don't contact anyone from your old life unless it's absolutely necessary. If you do need to do it, call from somewhere you do not stay, such as a different state.

Use a burner phone (a prepaid phone that you don't need to show ID to buy). Keep the call under three minutes, and say nothing that will give away your location or plans.

Once finished, remove the battery and SIM card from the phone and smash it.

Related Chapters:

- Traveling

REMOVE TEMPTATION

If you are not an obvious target to begin with, you are less likely to become one.

An important concept for increasing safety in all situations is to "be the gray man."

The gray man/woman blends in. He/she is unremarkable, and does not:

- Act boisterously.
- Show off expensive or flashy clothes, jewelry, phones, or obvious identifiers, like tattoos.
- Act like a tourist (taking pictures, looking at maps, speaking a foreign language, etc.).

Another important aspect of not becoming a victim is physical fitness. If you look like you can fight off, run away from, or chase down someone, you're less likely to be a target.

Combining being a gray man, physical fitness, and awareness makes you a terrible mark. Most criminals won't bother, and will go after an easier target—of which there are many.

HIDE YOUR VALUABLES

Appearing as if you have nothing of value lessens the chance of your home or car being invaded for monetary gain.

When hiding stuff, consider the amount of access you need in comparison to the amount of security you need. The longer it takes to conceal an object, the longer it will take to uncover it, both for you and for the criminal.

On Your Person

When out and about, don't bring any unnecessary valuables. Carry a dummy wallet containing a little bit of cash, a single form of old ID (with an old address), and an expired credit card.

Keep spare cash and a credit card in a secret place, like:

- A secret pocket.
- The sole of your shoe
- The lining of your clothing.

To hide something in the sole of your shoe, hollow out the heel on the inside, under your insole. Pad the dead space and glue the insole back.

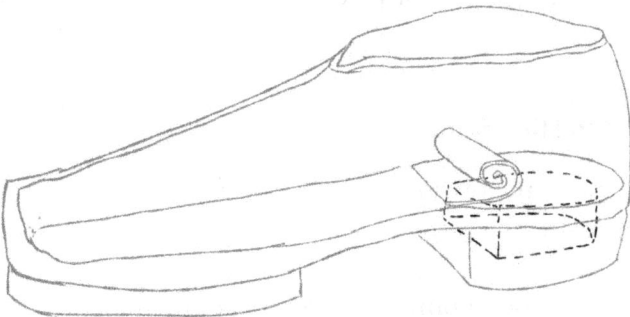

In Your Car

Don't leave anything valuable in sight . Even small change will attract thieves. At the very least, put it under a seat or in the glove box.

The trunk is the best place for anything of value, as it's out of sight and has a secure lock. For more intricate hiding places, try using the insides of the door panels or sewing valuables into the upholstery.

Don't leave obvious signs that you're a woman. Instead, leave signs that suggest you're a man, such as a cheap sports cap.

At Home

There are many good hiding spots for small to medium items in the home, but hiding household appliances, such as your big-screen TV, is not practical.

Close your blinds to prevent people from being able to see into your home, and don't leave anything valuable out in the open. This includes signs of new purchases, such as the packaging from your new game console.

Lock your car in the garage (which will also make your routine harder to track), and lock all your tools away.

You have several options for places to put anything you want to hide inside your home.

Where Not to Hide Stuff

Everyone knows the obvious hiding places, especially thieves. Do not hide things in the following places:

- The master bedroom. Leave your dummy wallet with $20 and some cheap jewelry as a decoy.
- A child's bedroom.

- Any cliched hiding spot, such as your underwear/sock drawer, mattress, freezer, or toilet tank.

Easy Hiding Spots

These hiding places are quick and easy to create and access with no damage done. They are also good for hotel rooms and office buildings.

When hiding something in a hotel room, don't forget to hang the "do not disturb" sign. Then place your item(s):

- In the bathtub spout. Pack it with toilet paper too so the item doesn't fall out.
- Inside a hollow shower-curtain rod.
- In the hems of window curtains.
- Under the ironing board cover.
- Taped to the bottom of a bottom drawer.
- Taped to the bottom of heavy furniture.
- Inside zippered cushions.
- In picture frames.
- Inside bathroom products. Waterproof the item in plastic first.
- In a fireproof safe bolted to the floor and hidden, but not in the master bedroom.

Medium Hiding Spots

You may need a tool to make/access these hiding spots:

- Inside the landline housing. Do not hide anything in an electrical socket.
- Underneath a bit of pulled-up carpet in the corner of a closet.
- Inside the TV housing.
- In a hollowed-out book. Use a razor to hollow out a few pages.

- In a hollowed-out can. Open the bottom (do not remove it) and replace the contents with your item(s).
- Inside venting. Lay the item on its side so it won't fall out of reach.

Hard Hiding Spots

You will need to do some minor construction to make these hiding spots.

- Inside dead space in a wall or door (behind the medicine cabinet, for example).
- In a safe bricked into a wall. Cover it with a painting.
- In hollowed-out furniture legs.
- Under a kitchen cabinet. Remove the kick board, hide your stuff, and use Velcro to stick it back.

Creating a Secret Room

You can make secret rooms from existing single-entry rooms or larger areas of dead space, such as the one under your staircase. To do this:

- Remove the door and molding.
- Fill the hole with drywall, except for a small crawl-in entry point.
- Create something to cover the entry that you can close from the inside, such as shelving that slides or pivots.

Your secret room can double as a safe room.

Related Chapters:

- Safe Rooms

PROTECT YOUR PRIVACY

The less people know about you, the less likely you are to be a victim.

Here are some tips for protecting your sensitive information:

- Never tag your keys with an address.
- Use a PO box for correspondence. For places that don't accept PO boxes, swap out "PO Box" for "Apt".
- Have coworkers screen calls and visitors.
- Destroy discarded mail.
- Remove your name and/or title from your reserved parking spot.
- Stay off all public lists, such as phone directories and school contact lists.
- Stay out of the media.
- Static interference with your phone, radio, or computer service can be an indication the device is bugged. Buy a cheap RF signal detector to check.

When people are fishing for information from you, use one or more of the following tactics.

- Be direct. Say, , "Sorry, I don't know."
- Ask a question back, such as "Why do you ask?"
- Change the topic.
- Refer the person to an alternative source. For example, you might say, "I think Bill knows about that."

Related Chapters:

- Digital Security

LOOK SECURE

Not being an obvious victim is good, but making it clear that your home is secure is better.

Leave signs that your home is not a good target. You want your house to look more secure than the rest of the houses on your street.

Do the following, even if it's all fake. Unless you're a specific target, criminals won't bother to find out. It's easier to go to the next home.

- Put security stickers on any windows near doorways and any sliding doors.
- Have a security sign in your front yard.
- Mount security cameras. If they're fake, make sure they have battery-operated red blinking lights.
- Place men's shoes and/or large dog bowls near the front and back doors.
- Have a "Trespassers Shot on Sight" and/or "Guard Dog on Site" type of sign.
- Remove signs of absence. Collect mail, empty bins, and remove any flyers or markings. Criminals use tape or signs ways to mark a home for potential robbery.
- If you're not home, use staggered timer lights that turn on in the kitchen and living room in the early evening, and in bedrooms at night.

SECURE ENTRY POINTS

The more secure your entry points are, the harder it is for a criminal to enter your domain.

Exterior Barriers

Your first line of defense is your front yard. You want one easy way for visitors to get to your front door, and nowhere else.

You can achieve this with funneling. Create an obvious path from the sidewalk to your front door, and put obstacles everywhere else.

The obstacles can be man-made (ponds, fencing, limestone/rock wall) or natural (hedges, prickly bushes, dense vegetation).

Garage

Always close and lock your garage. Padlock it if you are going away for more than a few days.

Keep your garage door opener somewhere safe, not visible in your car, and always lock the door between your house and garage.

A clever criminal can easily reach the manual pull string to open your garage door. Tie it up or cut it off.

Keys

Never hide your keys outside or leave them in your car, even if your car is in a secure garage. It is better to leave a spare set with a trusted neighbor instead. If you lose your keys, change your locks immediately.

Change all exterior locks when you move into a new house or apartment.

Windows

Install solid locks on all your windows, and use plexiglass or security film to make the windowpanes harder to smash.

Ensure any wall mounted air-conditioning units are secure so criminals can't pull them out and climb through the gap.

If you install security bars, ensure you still have ways to escape in case of a fire.

Placing pea gravel under your windows on the outside will allow you to hear the crunch if an intruder approaches at night.

Sliding Glass Doors

Sliding glass doors are notoriously easy to breach. A security screen is a must.

Consider installing a patio door bar (charley bar), which is a specially designed bar that goes across the door, making it jimmy-proof.

Exterior Doors

An exterior door is one that allows access from the outside into your home or garage. To make your home safer, do the following to your exterior doors.

- Install solid doors with strong locks.
- Use a deadbolt with a heavy-duty strike plate, and make sure it is installed properly, so the bolt goes all the way into the strike plate.
- Screw the hinges and locks in with three-inch wood screws, and use jam pins on the hinges so they can't be popped" out.
- Install security screens and remove all doggy doors.

You can also use these tips for your safe room.

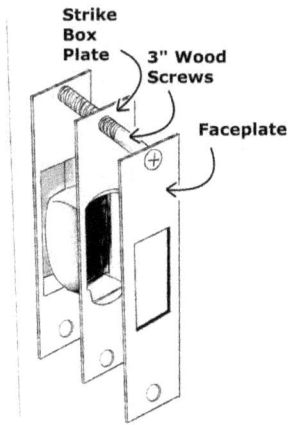

Hotel/Office/Public Building

Always use the deadbolt to lock your room. Chain and/or bar locks are only a backup.

Lever handle knobs are easy to defeat. Stuff a towel under the door to block the gap, or put one in the gap of the handle.

Barricades

How you barricade your doors depends on how they open.

For outward-opening doors, install an eyelet in the wall and run a strong cable from the eyelet to the handle. When the door is pulled, the cable will prevent it from opening.

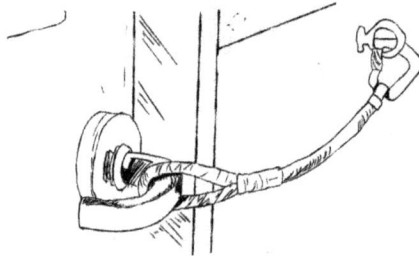

In a spur-of-the-moment situation, tie something from the doorknob to a fixed point (preferred) or heavy object. Power cords make good improvised rope.

For a fixed-point anchor, make the rope as tight as possible.

If it is a heavy but movable object, make sure that it will jam if pulled, so the door can't open.

The object doesn't have to be heavy. If it jams and is not easily broken, it will work. A broomstick fixed horizontally across the door frame and anchored tightly against the doorknob will not move easily.

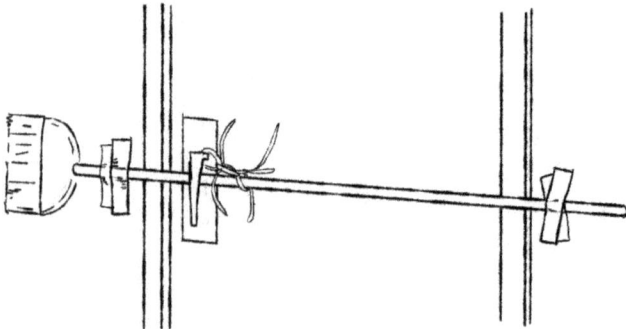

With inward-opening doors, install eyelets on both sides of the door and thread a pole through them to prevent the door being pushed open.

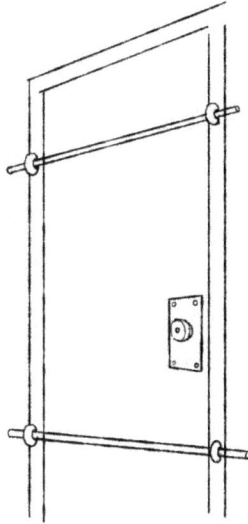

When that is not available, or for extra barriers in the case of an intruder, do as many of the following as you can:

- Barricade the door with heavy furniture.
- Place wedges between the frame and the door.
- Put wedges underneath the door handle.

INCREASE VISIBILITY

You want to be able to see as much as possible around your house from your windows and/or security cameras.

To do this, first figure out where you can't see. Look around during the day and at night for blind spots and potential hiding places, especially near entry points.

Once you have the blind spots identified, remove anything obstructing your view. You may need to trim foliage, for example.

Lighting helps you see and is a great deterrent. Install motion-sensor flood lights all around your property.

Security Cameras

Installing security cameras in and around your house act as a deterrent, and allows you to collect evidence. Make sure your security cameras are tamper-proof by placing them out of reach and in sturdy casing. Do this for your lighting as well.

A narrow focus (walkways or doors) is good for viewing faces, while wide-angle shots of your yard will capture vehicles.

Review the footage regularly (every Sunday morning, for example). You can also monitor your cameras live via an app on your smartphone. This is great for checking your internal feed from your room if you hear a noise, or if you have workers in your home while you are out.

DIY Listening Device

When you want to eavesdrop on people, use a mini hearing aid to amplify your hearing.

Alternatively, turn a set of headphones (or any speaker) into a listening device. Switch the positive (red) and negative (black) wires that go into the earbud. Plug the audio jack into a recording device or cell phone.

A voice-activated digital recorder will only record when people are actually talking.

If you want to listen to the conversation in real time, use a cell phone set to auto-answer. Turn it to silent and call it when you want to listen in.

INSTALL WARNING SYSTEMS

There are several ways to set up early warning systems that will alert you to intruders.

Motion-sensor lighting is a minimum precaution. Installing an alarm system is an option too. Make sure it is wireless and tamper-proof.

Neighborhood Watch

Creating and participating in a neighborhood watch program uses the "safety in numbers" principle and builds a trusted community. This has numerous benefits:

- Neighbors can warn each other of unusual activity.
- Your family will know which house(s) they can go to for safety if needed.
- It's easier to solve conflict between neighbors if you are on friendly terms.

Dogs

There are two types of dogs to consider for security purposes.

A watch dog is an alarm system. It will make a lot of noise at intruders while (usually) staying family-friendly.

A guard dog is also an alarm system, but is more likely to engage intruders. Most guard dog breeds are larger than watch dog breeds.

Either type of dog is a good choice, and you can choose a specific breed depending on what characteristics you want. Mixed breeds work well too, and often have fewer health issues.

An effective watch or guard dog doesn't need to be too big, but anything too small (such as toy dogs) will not deter most criminals.

Whatever you decide, train your dog(s) properly with positive rein-
forcement, and take barking seriously. Start with basic obedience
training (sit, stay, come, etc.).

All watch dogs will make noise at an intruder instinctively. Some
guard dogs may just sit and growl. Encourage them to investigate
strange noises instinctively and by command (, "Check it out") and
to bark when there are visitors. Train your dog to stop barking on
command as well.

Depending on your property, taking your dog on a twice-daily walk
around your boundary line is a good idea. Eventually he will patrol
it on his own through-out the day.

Many dogs will not protect you instinctively. Train yours to attack on
command, and not before. He must also stop attacking on
command.

Your dog's loyalty will come from being shown love and discipline.
Treat him well, and he will be more likely to risk himself for you.

Trip Alarms

Trip alarms are good to set up anywhere you think an intruder
might approach, and/or in places you think are not secure enough,
such as in dark spots and shed entrances or across windows and
fence lines.

To set up a trip alarm all you need is some fishing line and a cheap
pull-pin-style panic alarm. Ensure the alarm is loud and waterproof.

For ground-level trip alarms, attach the alarm to a tree (or whatever)
at about shin height. Tie fishing line from the pin to another tree
across the path you want to secure. It needs to be taut, but not too
tight, or you'll be more likely to get false alarms.

Panic Phrase

Have a family panic phrase to communicate to each other when something is unsafe and/or help is needed without it being obvious. For example, if there is an intruder in the house, you can use the panic phrase to let family members know not to come home and to call for help.

PLANNING AND PREPARATION

A plan is a predefined series of steps you will take to achieve a specified outcome.

Preparation is using the information from your plan to get as ready as possible before you act.

In almost all areas of life, planning and preparation increase the chances of success. In the context of the subjects discussed in this book, being successful means escaping danger.

CREATING PLANS

Making plans correctly in everyday life is the best way to internalize the process. That way, when you have to make a plan under stress, you can.

When going through these steps, access the situation objectively. Rely on facts first and past experiences second.

When there is no time, follow the process as best you can. The human mind is amazing, and you can make intelligent calculations very quickly.

Decide Your Goal

Without a clear aim, you won't be able to imagine the best way to achieve it.

Evaluate Strengths and Weaknesses

Evaluate your strengths and weaknesses, along with those of your team and your enemy (if applicable).

Consider:

- Skills.
- Resources you have, such as tools, weapons, and people.
- Resources you need.
- Obstacles that are known, likely, and/or possible.

Formulate Several Possible Plans

Creating more than one plan prevents tunnel vision. It also gives you backup plans. There isn't always time to create more than one plan, but if there is, do it.

Predict Outcomes

Predict the outcomes of each plan, considering the pros and cons. Cons must include any possible negative consequences.

Prioritize Your Plans

Choose what you believe to be the best plan based on its chance of success. Simple plans usually mean fewer things that can go wrong. Choose two backup plans in order of preference as well.

Analyze Your Plans

Analyze each of your chosen plans in detail. Practice them if circumstances allow it. Consider all possible things that may go wrong, and double-check the details.

PREPARATION

Once you have finalized your plans, communicate them to whoever needs to know (your family members, for example) and begin preparations. Preparations include gathering resources and rehearsing scenarios.

Gathering Resources

To gather resources, write a list of all the things you need to carry out the plan effectively and how to acquire them. Once you have your list, go out and get the things on it.

Rehearsing

Rehearsing is practicing the plan in real time. It ingrains the necessary actions in your mind, which will make it easier to carry out the plan in times of stress. Try to create as close to a real-life scenario as possible. This will also help you uncover and fix any flaws in your plan.

It's not unlikely you'll have to operate in the dark at some point. For example, there might be a power outage at night, or your kidnappers might blindfold you. Rehearse for this. Close your eyes, wear a blindfold, train at night with the lights off—whatever you prefer.

The ability to navigate your home in the dark is vital, as it will give you an advantage over any intruders. Keeping things in their place and general home tidiness help.

Be mindful of the need to keep your rehearsals safe. This shouldn't be too much of a problem, because if something is too dangerous to do in rehearsal, then it's probably not worth trying in real life either.

TRAINING

Another part of preparation is the general training of your mind and body.

Mind

Training your mind will keep you calmer in stressful situations. Do this with regular meditation.

Box breathing is a deep breathing method devised by Mark Divine. You can use it to calm yourself quickly during stressful situations, and/or as a form of breathing meditation.

- Empty your lungs of all air.
- Hold with your lungs empty for four seconds.
- Inhale through your nose for four seconds.
- Hold for four seconds.
- Exhale for four seconds.
- Repeat for as long as you need/want.

Body

Training your body will keep you physically strong. This will make you less of a target (predators prefer to prey on the weak) and boost your capability for fight or flight.

To train your body, you must eat well and exercise.

Eat a balanced diet, including a lot of vegetables. If you want an in-depth eating plan, visit:

www.SurvivalFitnessPlan.com/Nutrition-Guidelines

For physical training, exercise using skills that will aid you in fight or flight, such as, self-defense and/or parkour.

As a basic training routine, repeat the following five times. Do it at least three times a week:

- Thirty seconds of aggressive non-stop punching on a punching bag.
- A 60-second sprint.
- Thirty seconds of rest.

KEEP HANDY ITEMS CLOSE BY

Having a few things within reach access can make a big difference. At a minimum, keep the following items in spots you're in often, such as your bedroom, car, or office.

- Flashlight.
- Phone and charger.
- Weapon (gun, knife, steel-barreled pen, pepper spray, baseball bat).

COVERT SURVIVAL KIT

A covert survival kit is a bunch of items you can use for escape and survival, which you spread out and hide around your body. You do this to give yourself a better chance of retaining those items if you're searched.

The best places to hide things are where people won't want to search, like in your pubic hair, cavities, or fake wounds.

Other possibilities include:

- Shoes (tongue, sole).
- Clothes hems.
- Waistband.
- Hair.

Consider whether an item needs to be accessible when your hands are restrained. At a minimum, include:

- A button compass.
- Cash.
- An LED flashlight.
- Paracord.
- Paperclips.
- A poncho.
- A smartphone.
- A "tactical" pen.
- * A Ferro toggle.
- * A lighter.
- * A razor blade.

Some additional items to consider are:

- Bobby pins.
- Food.

- A handcuff key.
- A local map.
- Water purification tablets.
- *A knife.

*These items may not get through secure areas, but they're cheap, so if they get confiscated, or you throw them away beforehand, it's no big deal.

Button Compass

A compass will make navigation much easier, but most button compasses are inaccurate. Ensure you get a high-quality one. Silva and Suunto are reputable brands.

Cash

US dollars are the preferred currency to carry, besides the local currency. British pounds or euros are also widely accepted. Hide some smaller bills and have a dummy wallet that your captors can confiscate.

LED Flashlight

A small LED flashlight can help you see in the dark, signal for help, and attract fish in a survival situation.

Paracord

Replace your shoelaces with paracord, and use it for cutting through restraints, fishing, repairing stuff, and more.

Paperclips

Carry several larger, heavy-duty paperclips in your pocket and/or clipped to your clothing. They are great for picking locks, as well as for survival. You can make improvised fishing hooks out of them, for example.

Poncho

A clear plastic poncho is useful for shelter, water collection and more. Unfortunately, it is not practical to carry one around unless you have a backpack.

Smartphone

The modern smartphone is the ultimate escape and survival tool, until the battery runs out. It will also be the first thing to get confiscated. Some things you can do with it include:

- Call for help.
- Navigate with the built-in compass and/or GPS.
- Use it as a flashlight.
- Take notes and photos.
- Keep some ebooks, such as survival and first aid guides, on it.
- Use it as an improvised signal mirror.
- Start a fire with the battery. Only do this if there is no other use for it.

"Tactical" Pen

The best type of tactical pen is one that you'll carry. Any simple stainless-steel pen will work. Look for one with the following characteristics:

- It's refillable.
- It writes well.
- It has a clip.
- It has a flat top.
- It's easy to replace/inexpensive.

- It can pass as a normal pen.

Most of the tactical pens on the market do not fill these requirements, especially the last one. A few of the ones that do are:

- Zebra 701.
- Zebra 402.
- Parker Jotter.
- Fisher Space Military Pen (this one is a little more expensive, but still under $20).

Ferro Toggle

A ferro toggle will help you start a fire in an emergency.

Be sure you get a ferro one, rather than a flint or magnesium, so it is easier to create a spark without the special striker.

A toggle shape is not as easy to use as the more common rod type, but you can attach it to clothing (as a zipper toggle, for example), which makes it less likely to be confiscated.

Lighter

As long as it doesn't get wet, it is easier to start a fire with a lighter than a ferro toggle.

You can also use it as an improvised diversion explosive, to signal for help, and for self-defense.

Razor Blade

A razor blade is the next best thing to a knife, and is easier to hide.

Bobby Pins

Bobby pins make good improvised lock picks, and work better than paperclips on certain locks.

Food

A high-calorie nutrition bar can go a long way when you are stranded.

Handcuff Key

Handcuff keys are easy to hide and make escaping from handcuffs much easier. Depending on where you are, it may be illegal for you to carry one.

Local Map

This is useful for navigation and as writing paper if you're desperate. Never deface a map to the point it can't be used.

Water Purification Tablets

Drinking water is essential for survival, but drinking contaminated water will make you sick (or worse). To avoid that possibility, carry water purification tablets, which are small, easy to use, and reliable.

Knife

A good knife is by far the best escape, evasion, and survival tool there is. A multitool or pocket knife is not as good, but better than nothing—and still very useful.

Related Chapters:

- Picking Locks

BUG OUT BAGS (BOBS)

A bug-out bag (BOB) is a single bag of supplies you can quickly grab and go when needed. It's basically a survival kit with at least several days of provisions. It must have the ability to provide you with water, food, shelter/warmth, fire, rescue, health, and security. Many items in it will be of a general nature, but when you pack it, you should also consider likely events in your area. This way, no matter what the emergency, you can grab your BOB (if it is safe to do so) and bug out.

Everyone in your household, including your pets, should have their own BOB, and they should keep it somewhere easy to access in case of an emergency. Under the bed or next to the nightstand are good options.

Assign responsibility for pets, infants, etc. and their BOBs. Do it now, so there is no confusion when an emergency arises.

What to Put in Your BOB

The exact contents of your bag will depend on what you're comfortable using and what events you feel are most likely to happen. You can also add some personal and/or comfort items if you have the room and weight tolerance (you may have to carry it all day, every day). The bag itself must be comfortable and sturdy.

Once you have put your BOB together, ensure you rotate the perishables every few months.

Here is a list of items to consider including in your BOB:

- Cash (small bills).
- Knife (steel).
- Multitool.
- One liter of water (minimum).
- Water filter (portable/hiking style).

- Food (long-lasting and ready to eat; think energy bars, trail mix, multivitamins and electrolyte mixes).
- A spare set of clothing.
- Emergency blanket.
- Poncho (transparent white is best).
- Lighters.
- Ferro rod.
- Flashlight (headlamp).
- Whistle.
- Shortwave radio with AM/FM (battery-operated and compact).
- Batteries.
- GPS-capable cell phone (with SIM card and charger; a cheap "burner" phone is ideal).
- Maps.
- Compass.
- First aid kit (with antibiotics).
- Toiletries (essentials).
- Sewing kit.
- Duct tape.
- Paracord (5m).
- Weapon and ammo (if legal).
- Notebook and pens/pencils.
- Plastic bags.
- Photocopies of important documents (see end of this chapter).
- Swimming goggles.
- P100 mask with an air vent.
- Special-needs items.

For infants:

- Food/formula.
- Water.
- Clothes.
- Comfort toys/blankets.

For pets:

- Food.
- Water.
- Leash.
- Toy.

It is a good idea to get a cage for your pet and train him/her to sleep in it. That way, it will be comfortable for him/her to stay in when you need to leave in a hurry. Keep his/her **BOB** on top of the cage.

CACHES

A cache is a hidden store of supplies.

You may have caches in your home, at rally points, along your routes to bug-out locations, or anywhere else you think makes sense.

You can also have different caches for different things, either to separate items or to pack them for specific scenarios.

Containers

The container you choose for your cache must protect the items you are storing. It must be waterproof, airtight, and corrosion-resistant. Other characteristics to consider depend on how easy to access it needs to be and where you're going to hide it. For example, is it suitable for burial?

A PVC pipe with sealed ends is a popular option, as it is durable, inexpensive, and easy to waterproof, but any other durable box will work as long as you seal it properly. If it has a rubber lining, that will make your job easier. Test the seals by submerging the cache in hot water and looking for bubbles.

Additional Protection

Waterproof the individual items before putting them in the waterproof cache. You can use heavy-duty trash bags, vacuum sealing, plastic sheeting and duct tape, etc. Before sealing the items in, add desiccants and remove as much air as you can.

Adding desiccants will absorb extra moisture. Silica gel packets are common and cheap. Use 5g for every 3.5L (1gal) of space. If in doubt, add more.

There are many other choices for desiccants, which may or may not work as well. These include rice, salt, zeolites, calcium sulfate, and kitty litter.

Hiding Your Cache

A big factor in deciding where to hide your cache is accessibility. You need to be able to access it in an emergency, as well as for maintenance.

Another factor is concealment. Put the cache somewhere that is not obvious, but that is easy for you to relocate. Burying your cache is a good option, especially if it is off your property. If you need semi-regular access, consider a shallow bury—for example, place it in a small depression under a large rock.

When the cache is on your property, you can hide it in your walls or roof.

Other options include hiding it at your workplace, in a storage container, in a PO box, on a rooftop, or even underwater (if you have a boat moored at the local harbor, for instance).

Some places to avoid include:

- Private property that's not yours (unless you are paying for it, in which case, keep anonymous if possible and never miss a payment).
- Populated places (parks, beaches, vehicle access roads).
- Abandoned buildings.
- Anywhere with security cameras.
- Places that may be developed in the future (outside urban areas).

The way you store your cache will also determine its location. For example, if you're burying it, you'll want to avoid choosing ground that contains obstructions, such as rocks, large tree roots, or pipes.

You'll also want to avoid ground that is high in moisture or prone to rain run-off. In general, don't bury it in lowlands.

Wherever you choose, you need to scout the location out before actually putting your cache there. Decide on a possible area from home first using Google Maps/Earth. Then go out there to assess it further. Check out exactly where you think you will stash/bury your cache, as well as how secure the area is.

You will need to get the cache and tools out there and have enough time to stash (or bury) it without anyone seeing you. Stake it out at different times, too, in case there is a change in activity level on weekends vs weekdays, or at night vs during the day.

Once you have an exact location, you need to remember where it is. Perhaps you can remember without a prompt, but I wouldn't rely on that alone unless you have a photographic memory. Things (especially your memories) change over time. A better idea is to write non-specific instructions that you understand, but that will be useless to others. Other options are to store the location in your GPS, record the grid references on a map, and/or include a small Bluetooth tracker in the cache.

Keeping the Secret

There is no point hiding a cache if other people know about it. In fact, don't even tell anyone you plan to do it. If you live in a rural area where word spreads easily, purchase supplies in from a different town.

When physically hiding (or accessing) your cache, you need to be as covert as possible. Do it at dusk or dawn on a Sunday or Monday, and wear gloves so there are no fingerprints. Use a flashlight only if you need to, and make sure it's red or blue (never use white light). Ensure you leave no signs of your presence. This means parking your car out of the way and hiking in without making an obvious trail. Unless you're burying your cache, you also need to consider ways to camouflage it.

Ensure no GPS devices (phones, cars, etc.) record where you're going and have a cover story in case anyone comes along. For example, say you're doing a time capsule project or treasure-hunting with a metal detector. Take equipment to confirm your cover, and ensure you have food and water.

If you need to access your cache, take the same precautions. Always use a different path in/out (to prevent making trails) and minimize access to it. The more often you access your cache, the less secure it is. To improve security, you can also create decoys and/or misdirections by burying a layer of trash above the cache.

Car Supplies

You can store additional supplies in your car. Keep them in the trunk for security, except for the last two items, which you'll need to have handy in case of an emergency.

- Blankets.
- Additional food, water, flashlights, and batteries.
- Fuel.
- Recovery and repair supplies.
- Entertainment (books, cards, laptops, etc.).
- Chargers.
- A small fire extinguisher.
- A glass-breaker.

Do not put your personal BOBs in the trunk. Keep them within reach in case you need to leave your car in a hurry.

Important Documentation

Gather all the following documentation. Keep the originals in a fireproof safe (or some other secure place) and tell your family its location. Photocopy everything and keep the photocopies in your BOB. Ensure everything is kept current.

- Your will.
- Your powers of attorney.
- Emergency/important contacts (numbers and addresses).

- Your passport (or other ID if you don't have one).
- Insurance information.
- Proof of residence (utility bill).
- Access to finances (do not keep a photocopy of this in your BOB).
- Personal info sheet and recording.

A personal info sheet is a single sheet that will aid rescuers in finding and/or identifying you. Each family member should handwrite their own info sheet and make an audio recording of the information. This is so rescuers will have writing and voice samples.

Each sheet/recording should include the following:

- Name.
- Nicknames.
- Place of birth.
- Date of birth.
- Address.
- Phone number.
- Physical Description (including specific identifiers like tattoos or birthmarks).
- Prescriptions (eyes, medication).
- Instructions for prescriptions.
- Vehicle (color, type, license plate number).
- School/work address and contacts.
- The contact details of closest friends/relatives.
- Hobbies.
- Education.

Related Chapters:

- Rally Points

RALLY POINTS

A rally point is any pre-designated place for your party to meet up in case something goes wrong. It isn't strictly an "item,", but it is handy to "keep."

There are a few different types of rally points, and it is common to have different rally points for different situations. If you do have several rally points, plan for when and how to use each one.

Consider stashing some basic supplies, such as food, water, and flashlights, at your permanent rally points.

Never share the location of your rally points with outsiders.

Temporary Rally Points

Assign a temporary rally point anytime you are somewhere new. Make it an easy to find place, such as a landmark. Most people do this anyway, saying things like, "If we get split up, meet at the mall entrance at 3:30" or "If you get lost in supermarket, go to the #6 checkout."

Primary Rally Point

This rally point is where you can meet up once you escape an incident, such as a fire or a home invasion. Make it somewhere relatively close and safe, like a trusted neighbor's house or a local 24-hour gas station. Decide when to go to the rally point, as well as:

- How long to wait there before going to your secondary rally point.
- When to skip it and go straight to the secondary rally point.

Secondary Rally Point

This is an alternative rally point that you can go to when the primary rally point is not feasible. It should be somewhere public, but not a place it's obvious you would go—, a pub you never frequent, for example. It must also be easy to get to from common places such as home, work, or school.

Hideouts

Hideouts are not really rally points, because you will stay in them for an extended period.

In most cases, you will meet your family at a rally point and then make your way to your hideout, but you could also meet at the hideout.

You may want to stockpile food, water, a first aid kit, and other supplies at your hideout.

Examples of good hideouts include:

- Abandoned buildings that you have concluded are safe.
- A hotel room (though you can't stock supplies there).
- A "secret" property out of the city or in a neighboring town.

Routes

You need to plan several entry and escape routes to and from all types of rally points, as well as to consider the best times to come and go without raising suspicion. These will depend on the situation.

Related Chapters:

- Home Invasion

EVASION PLANS

This section contains a selection of plans to follow in various situations, along with additional information. Use the plans as they are, or adjust them to suit your needs.

GENERAL EMERGENCY ESCAPE PLAN

Plan for escape every time you enter a new space. Determine:

- Three things you could use as a weapon.
- Where the exit points are and which ones you will use. Designate a primary and back-up.
- Temporary rally points (if you're in a group).

Related Chapters:

- Rally Points

CALLING EMERGENCY SERVICES

Know the emergency numbers of the country you are in.

Ensure children can reach the phone and know how to use it in case of an emergency. Leave a list of emergency numbers near it.

When you call emergency services, talk clearly and slowly, using the following format:

- I need (insert emergency service) at (location).
- My phone number is (optional, but recommended so they can call you back if needed).
- Describe the incident and give any additional pertinent information, such as a description of the victim and/or culprit, details of injuries, or the next of kin's phone number.

Don't hang up until you're told to, in case emergency personnel need to give you instructions.

When you can't talk, call and leave the phone off the hook so they can listen in. Tap SOS on the speaker if you can. Even if there is dead silence, they may trace the call.

Another option is to send the information to all your contacts by mass text. Start your text with "This is not a joke. Send the police."

Make sure to put your phone on silent, in case one of your contacts calls you back.

To hide the fact you are calling the police, pretend you are speaking to someone else, such as your mum or your spouse.

Sound natural while answering the dispatcher's questions. Do this by answering the question directly and then adding improvised content. For example:

- **Dispatch:** What is the emergency?

- **You:** Hi, honey. Just confirming we're meeting for dinner tonight.
- **Dispatch:** Do you require police assistance?
- **You:** Yes please, soon. I'm getting hungry already. I'm walking past (street name) at the moment, so I should be able to meet you at (name of place) in five or so minutes.
- **Dispatch:** Okay ma'am, we are tracking your cell. You can stop talking, but don't hang up.
- **You:** Okay, great; thanks.

If you call emergency services by accident, don't hang up or someone may be dispatched. Inform the operator it was a mistake.

FIGHTING BACK

When someone is trying to abduct you, your best chance of survival is to fight back and make as much commotion as possible. Once they get you in their vehicle, your chances of escape drop dramatically. Scream for help and attack your abductor in vulnerable areas.

- Eyes (gouge).
- Groin (grab and twist, kick, knee).
- Shins (kick).
- Fingers (twist).
- Throat/neck (elbow, poke).
- Piercings (rip them out.)

As soon as you're free, run towards "safe" areas (ones with good lighting and crowds). Knock things over as obstacles for him along the way.

Continue to scream for help as you run, and call emergency services. Set off car and shop alarms by hitting them or smashing windows.

If there are no safe areas and you find an unlocked car, get in it and lock yourself in. Blast the horn in an SOS pattern (... - - - ...).

Hiding under a parked car is a good last resort. Hold onto something on the underside of it, and kick if he tries to get you.

Fighting back is important, but so is knowing when to stop. When you know you are defeated your best chance of survival is to cooperate. This prevents further injury and/or restraints on yourself so you can take advantage of the next opportunity for escape.

To learn more about self-defense, visit:

www.SFNonfictionBooks.com/Self-Defense-Handbook

SEXUAL ASSAULT

The chances of getting killed during a sexual assault are higher than in an abduction for ransom. For this reason, only scream if you are likely to be heard; otherwise, your attacker may silence you.

Telling him you have an STD (herpes, hepatitis B, AIDS) may be enough to deter him. Be specific about what you have, so your story is more believable.

If that doesn't work, and even if you can't fight him off, do your best to get DNA samples (blood, skin, hair) to make him easier to catch after the incident.

After the sexual assault, is it important to preserve any evidence. Do not disturb the crime scene or wash yourself until instructed to do so by a medical examiner.

Go to a safe place as soon as possible (in case the attacker returns) and then call the police (or call them along the way if you have a phone). After calling the authorities, write down a description of your attacker. Timestamp and date it.

Once you have been "processed" by the authorities, seek counseling. Get a health screen three months after the incident to be sure you haven't acquired a delayed illness or disease.

Preventing Sex Offenses on Children

Teach kids the following to minimize the chances of them being sexually assaulted:

- That it's okay to say no to adults if they ask the kids to do something that you have taught them is wrong.
- To tell you if an adult asks them to keep a secret.
- That no one has the right to touch them anywhere that a bathing suit would cover.
- To tell you if anyone exposes their private parts.

- Not to loiter in bathrooms. (Always accompany them.)
- Not to approach, help, or accept things from strange adults.
- Not to enter other people's homes without your permission.
- How to use the panic phrase.

Related Chapters:

- Install Warning Systems

STALKERS

In this section, the terms "stalker" and "tail" both refer to someone (or numerous people) following you. This may be a traditional stalker (someone with an unhealthy obsession) or surveillance for a future crime. The best ways to prevent a stalker are awareness and randomization:

- Look around often.
- Check to make sure no one is following you when you leave a building.
- Change up your schedule when possible.
- Take different routes to places you go to regularly.

Recognition

If you notice the same people and/or cars repeatedly over significant time and/or distance, it is a sign of a stalker, but being able to recognize repeated sightings of unfamiliar people and/or vehicles takes practice.

Improve your ability to recognize people by mentally noting distinctive features: height, build, facial features, hair, how they walk, what they are carrying, etc. Checking their shoes is useful. Clothes are easily changed, but shoes are not.

Do the same with vehicles (make, model, size, color, license plate number, etc.). Take special note of illegally parked vehicles, parked vehicles with people inside, and people who look out of place.

Confirmation

When you think you have a stalker, take a few turns and see if he follows you. As a rule, if he is still following after three turns, you have a tail.

Check as you walk without being obvious by:

- Looking at reflections in mirrors, windows, and shiny objects.
- Making U-turns (to take an escalator, for example) so you can immediately see in the opposite direction.
- Leading him into a funnel, like a corridor or a highway. Be careful not to get isolated doing this, or you may be attacked.
- Going down dead ends (a cul-de-sac, for example).
- Slowing down your pace.

If you want to be more direct, turn and stare at him. An amateur stalker will get flustered and give himself away.

Action

Once you confirm you have a stalker, note down a description of the person/people and vehicle(s) involved. After that, you need to decide on what action you will take. You have two choices: confront or lose him.

Whichever one you choose, you must do it before going to your vehicle (if on foot) or arriving home. You want to deny him any information about you, and especially information about where you live.

Confront

This is a good option if you are in a public place where he is unlikely to attack you, and it is often enough to scare him off. Let him know that you know he is following you without directly accusing him of anything. Ask him the time, or say "Can I help you?" If he persists, be more direct. Tell him to stop bothering you in a loud, firm voice, so others can hear. Don't be afraid to press the emergency button if you're on public transport, or to alert the authorities.

Lose

When confrontation would be dangerous, or when you are unsure if the tail is real, try to lose him.

One of the easiest ways to do this is to go to a safe place, such as a cafe, library, or police station, and wait him out. On your way there, walk through heavily populated areas, as this may enable you to lose him naturally. An opportunistic criminal probably won't bother waiting around while you have a meal or read a book in a coffee shop. Make it obvious that you are settling in for a while.

You can also combine this with confrontation. If you sit down at a restaurant and just stare at your tail, he'll know that you see him. You could also take this opportunity to call a friend to meet you.

If he is still waiting when you leave, take a few quick turns to lose him. Another option is to enter a building with multiple exits and then leave from a different exit.

When in a car, drive through an area with many lights and/or stop signs.

A quick change of appearance will help you lose dedicated tails. Make one as soon as your tail loses sight of you momentarily, such as when you turn a corner or walk into a crowd.

Here are some ideas:

- Cover your face, with a hat and sunglasses, dust mask, or hoodie.
- Take off or put on a coat to display different colors and/or patterns.
- Put on shoes, a bag, and/or accessories.
- Change your posture.

High-Level Threats

The information in this section is for when you're dealing with a long-term stalker such as an ex-boyfriend, or ongoing harassment from an individual or group.

- Mix up your routine and behavior.
- Run surveillance on the threat, whether it is an individual or group. Find out everything you can (without turning into a stalker yourself).
- Increase security and awareness.
- Break all contact with the stalker, and ask family and friends to do the same.
- Do not visit the threat in person or escalate the tension in any way.
- Let people know what's going on (friends, family, co-workers, police). Keep them posted on your plans/itinerary.
- Gather evidence. Take screenshots of his phone number, make voicemail recordings, and keep written records with dates and times.
- Consider a restraining order, though this might make things worse if the person is unstable.
- Consider moving and/or disappearing permanently.

Malicious Phone Calls

The best course of action when dealing with malicious phone calls is to ignore them. Hang up and block the number. If they persist, or if there are threats of violence, start keeping records of interactions, call the police, and notify your phone company. Never admit to the caller that you're alone.

Related Chapters:

- Disappear Permanently

HOME SECURITY ROUTINE

This home security routine ensures your house is as safe as possible from intruders, fires, and other potential disasters. Do it before going to bed or leaving the house empty.

- Close and lock all doors and windows (including the garage).
- Close all blinds.
- Turn off interior lights.
- Turn on exterior lights.
- Unplug power bars.
- Ensure any gas appliances are off.
- Turn on the house alarm.

When everyone implements the security routine, it makes it easy to know if someone is in your home when you get there. If a window is open or there is a strange car in the driveway, that's a sign of a possible intrusion.

If no one is supposed to be there, then do not enter your house. Contact all your family members to see if anyone is home unexpectedly; if no one is, call the police and wait at a trusted neighbor's place for them to arrive.

ANSWERING DOORS

You should never trust a visitor without first vetting them. Even a trusted friend may be unwilling bait. Use the following tips for answering the door safely.

Get eyes on the visitor without unlocking the door. Use a peephole, security screen, or window. If it is a stranger, talk to him through the closed door/window. You want to let him know you are home, but never mention you are alone.

Beware of impersonators. Check for the following:

- Company ID.
- Uniform with a company logo.
- Company vehicle.
- Package tracking.

If in doubt, call the company to confirm.

Never open the door at night unless you have positively identified expected visitors.

When opening a door for a strange visitor, put all your weight behind it in case he suddenly tries to barge in.

Do not let unapproved people into your house. Under normal circumstances, this rule even applies to the police, unless they have a search warrant.

To minimize the need to open the door, request that all deliveries require no signature. Give instructions regarding where a delivery person can leave a package, or opt to pick it up from the store or post office yourself.

HOME INVASION

When one (or more) of your warning systems are activated, you and all the other members of your household must take immediate action:

- Grab your BOB (if possible).
- Head to the safe room.

Designated people may have additional roles. For example, they may be responsible for:

- Calling the police.
- Getting a weapon.
- Helping others (young children, the elderly, the disabled) get to the safe room.
- Clearing the house.

If you don't have a safe room, run out the door opposite the one the intruder entered through . Get to a safe place and call the police.

Never call out "Who's there?" It lets the intruder know that you're alone.

If you wake up to an intruder in your bedroom, pretend to remain asleep to avoid a violent confrontation.

Clearing Your House

To clear your home, you need a flashlight and a weapon. Do not try to do it without these things.

Adopt a defensive position. This is a place where you go to put yourself between the potential intruder(s) and your family and ensure no one gets past you. The top of the staircase works well.

Yell a warning, like "The police have been called and I have a gun."

Once you think the intruder(s) has/have left, you can clear the house. Be slow, quiet, and very careful.

Leave the lights off. The darkness gives you an advantage, since you know the layout of your home. Use your flashlight if needed.

Clear rooms one by one. Check behind all furniture and other hiding spots. Check behind yourself often.

Clear corners by "slicing the pie" This means clearing what you can see, then a little more, then a little more, etc. You gradually move around the "outside of the pie," clearing each slice as you go.

With each side step, scan from the floor up.

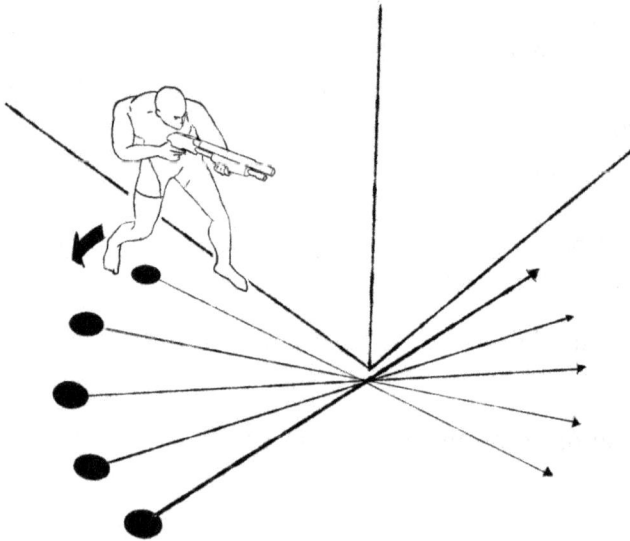

To clear a doorway, slice the pie as much as possible before entering.

Move through the doorway quickly to avoid the "fatal funnel" where you are most likely to get shot. Once through it, step out to either side with your back to the wall so you can clear the corners of the room.

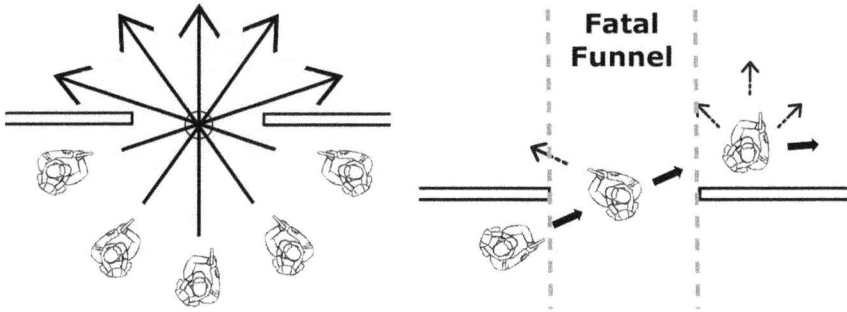

To clear a T hallway, clear one side at a time by slicing the pie.

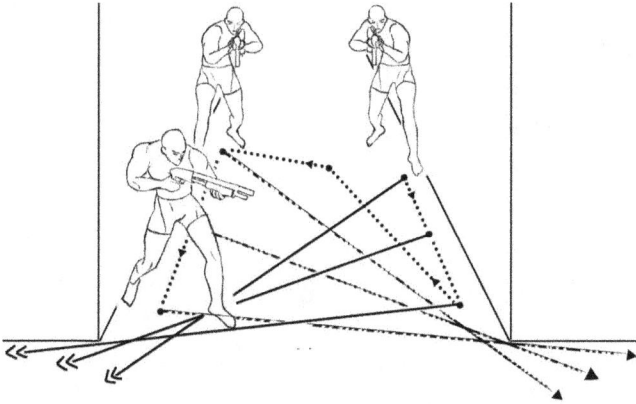

If you find an intruder, only attempt to restrain him if you have others to support you.

Military-Style Raid

When the government or well-organized criminals raid your home, unless you can make a quick escape, your best chance of survival is compliance. Stand still, with your hands up, and do as instructed. Do not volunteer any information.

Related Chapters:

- Restraining a Guard

SUSPICIOUS MAIL

A random mail bomb or biological threat is unlikely. Consider whether your lifestyle or career makes you a target.

Even if you aren't a high-value target, it doesn't hurt to be cautious. Look for the following signs of suspect mail:

- No return address.
- Unusual size, shape, weight, or texture (sticky, powdery, etc.).
- An excessive number of postage stamps.
- Smell.
- Wires or strings.
- Messy handwriting/spelling errors.
- Leakage.

If you identify suspicious mail, take the following action:

- Do not shake the package.
- Cover it, but do not touch any spillage.
- Put it in a plastic bag to prevent spillage.
- Leave and secure the room.
- Wash your hands with soap and water.
- Contact the authorities.
- List anyone who was in the room and give the list to the authorities (health and law).

ELEVATORS

Being alone in an elevator puts you in a vulnerable place due to the isolation. Stand close to the door and the buttons. When someone you have a bad feeling about gets on, step out quickly. If you are attacked, do not press the emergency stop button. Push all the floor buttons instead. Scream for help and try to escape as soon as the door opens.

CAR JACK

Here is how to give yourself the best chance of staying safe during a carjacking. A lot of this advice will also prevent general car theft.

In Your Car

You're generally safe while you're moving. It's when you stop or slow down that you need to increase your awareness levels.

Always have an escape route. To ensure you have enough room to drive out, leave enough of a gap so you can see the tires of the vehicle in front of you. Do not hesitate to drive away if you need to, even if you have to drive off the road.

Keep your windows up, doors locked, and the car gear in drive. If someone approaches your car, talk through the window. If you have to open it, as when a police officer asks you to, only open it a tiny bit.

When waiting in a stationary vehicle (in traffic, for example) check your mirrors frequently for anyone approaching. If you're a woman, keep a man's hat in the car and wear it if you need to wait alone at night (for someone to fix a mechanical problem, for instance).

Parking

Carjacking often happens when you're returning to a parked car. To prevent this, reverse park in well-lit areas away from possible hiding spots and close to the building exit or car park elevators.

Anti-theft devices (alarm, immobilizer, steering lock, tracker) are good additional deterrents.

Never park where you could be clamped or towed. If you are, don't wait outside. If nothing else, lock yourself in your car.

Never exit your vehicle without your keys.

When approaching your parked vehicle, keep your keys in your hand, with the keys pointing out through your fingers. This is a good improvised weapon and having your keys out makes getting into your car faster.

If your car has a push-button unlocking system, don't use it until you're ready to get in.

If there is someone suspicious near your car, do a U-turn and walk to safety. Ask for an escort (e.g., security, stock boy), and/or activate your remote alarm from a distance to encourage the person to move away.

Get inside and secure your car quickly. Once in a locked car, you can organize yourself and/or your kids, but do not linger.

If your car has been broken into, check underneath and inside it before entering, in case someone is hiding there.

If Attacked

As a rule, and especially when dealing with an armed thief, it is best to give up your car, but not yourself or other passengers.

If you have children in the car, tell (don't ask) the carjackers that you're getting them, and do it before exiting the vehicle.

When someone demands your keys while you are outside your car, throw them in the opposite direction of your escape route and run when he goes for them.

If he does not go for the keys, you'll know that you, and not your car, are his real target . If he has a gun, keep the car between you and him. Alternatively, run to the closest obstacle/barrier (a concrete pillar, for instance). Run from cover to cover until you reach safety.

If you're forced into a car, consider it an abduction.

When you're forced to drive, you can:

- Run red lights, honking your horn to attract attention.
- Drive to a police station.
- Get in a minor accident.

When you're inside the car and someone is trying to get in (and you can't drive off), hit the horn in the SOS pattern (... - - - ...) and call for help.

If someone sticks a weapon through your window, trap it on the dashboard and drive away.

Related Chapters:

- Elevators

CAR ACCIDENTS

This advice assumes you are involved in a legitimate car accident and not a collision scam.

When in an isolated area, it's best to keep driving until you get to a safe place, assuming your car is still safe to drive.

If you're already in a populated area, move the vehicle close to the scene of the accident, but not to a spot where it will obstruct traffic. Secure the area by checking there are no hazards, applying first aid, and warning oncoming traffic of the accident.

Call emergency services, and then take photos and notes. Record the date, time, weather, nature of the accident, etc. Date and time-stamp your notes once you finish writing, and sign them.

Exchange information with the other driver. Get his name, address, phone number, driver's license number, the name of his insurance company, and the insurance policy number.

Record the names, addresses, and phone numbers of any passengers and/or witnesses. If the vehicle doesn't belong to the driver, get the owner's details too.

Never admit responsibility for an accident. Do not sign any papers, agree to pay for damages, or downplay any injuries. Get a lawyer if you need to.

When it comes time to tow your vehicle, agree on a price and towing location before the towing is done.

Inform your insurance company of the incident, and follow up with a written report that includes copies of your notes, photos, and the police report.

See a doctor as soon after the accident as is practical (within 48 hours), even if you think you have no injuries.

Do not settle any insurance claims until you know the full extent of any injuries and vehicle damage.

Related Chapters:

- Common Scams and Petty Theft

VEHICLE BOMBS

Coming in contact with a vehicle bomb in your personal car is unlikely unless you are a specific target.

If you do think you are a target, your best defense is to check your car every time you get in it. This is also the best way to prevent vehicle surveillance via tracking devices, etc.

Keeping your vehicle dirty will make spotting any tampering easier. An additional (or alternative) way to catch that is by placing clear tape on your trunk, bonnet, gas tank, etc.

To assess a vehicle for a possible bomb, start by examining the exterior of it. Look for anything unusual, such as wires or any open doors. Search all around it, including in the wheel arches, bumpers, underneath it, etc. Give special attention to the area under the driver's seat.

After accessing the outside, look inside through the windows. Look for any suspicious items that weren't there before.

Finally, enter your car and check inside the glove box, under the seats, in the trunk, and any other spots you couldn't see from the outside.

To be extra safe, get a field-strength meter from your local electronics store. This device will detect radio transmissions.

GOING OUT

Whether you are at home or away, there are some fundamental steps you can take to keep yourself safe.

Commit the following information to memory wherever you go:

- At least two emergency contact numbers (parent, spouse, sibling).
- Your home/hotel address.
- The emergency services number (e.g., 911).

Survey the general area around your home or whatever new place you stay in for a night or more. Take note of:

- Exits from the area.
- Chokepoints in traffic.
- Police stations.
- Hospitals.
- Pharmacies.
- Water sources.
- Your country's embassy.
- Rally points.

Return to the area often, so you are aware of any changes, such as roadworks.

When you go out, dress practically and take only what you need. This will make you less of a target and give you more mobility for flight and/or fight.

Share your intended movements with someone responsible. Tell him when and how you will check in, and what action to take if you don't.

For example, establish that you'll send a text message every hour, and if you don't do so for three hours or more, he should inform your brother.

Related Chapters:

- Rally Points

SEARCH AND RESCUE

Knowing about search and rescue gives you foresight into how emergency services may try to find you.

When someone goes missing, make a plan and start searching ASAP. The longer the victim is missing the harder they will be to find, but searching without organization is often worse than doing nothing.

To make search and rescue easier, set the boundaries of where people can go. This includes creating and sticking to itineraries and planned routes. Doing this gives you a defined area to use for your primary search.

Identity Kit

An identity kit is a single sheet of paper with information about the missing person. You can give this to authorities/rescue team members. Make one for each family member, and include the following:

- An up-to-date color photo.
- Fingerprints.
- Stats (name, age, birth date, physical description).
- Medical conditions.
- A lock of hair in a sealed baggie.

Planning a Search

Establish one search leader and a command post. Set up the command post near or in the location to which the person may return (a camp-site, home). The leader will stay in the command post along with first aid supplies and a first aid caregiver, who can also be the search leader.

While the command post is getting set up, establish a primary search area and search teams of two to three people (if resources allow).

Divide the search area into sections, and assign each team to a section. Ensure each team has navigation aids and communication devices.

Each team will search their section and then report back to the command post so the leader can assign them a new section, call them back in, or give other instructions.

Search the most likely places first:

- Last known position.
- Most probable routes.
- Boundary limits.

Expand the search area as needed.

When making the search plan, consider:

- The missing person's strengths, weaknesses, demeanor, routines, health, age, etc.
- The weather.
- Available equipment (communications, first aid kits, food, water, navigation tools, signal flares, shelter, etc.).
- The strengths and weaknesses of search team members. Create pairs that complement each other's strengths. Make sure there's at least one first aid provider per team.

Searching

When carrying out the search in individual teams, use light and sound to attract the lost person's attention. Call out his name and use whistle blasts and/or flashes of light.

In the wilderness, use one person to guide a team along a prominent feature (stream, trail, etc.) while the others search deeper. Always keep in sight or hearing of each other.

Look in hiding spots (especially when searching for kids or kidnapping victims), and keep in mind that the missing person may be unconscious.

Rescue

When you find the missing person, inform the command post and apply first aid.

Give the victim food and water as needed, and then do as instructed by the search leader.

Consider setting up a shelter if the weather is bad or you need to wait for help.

Related Chapters:

- Tracking

TRACKING

Knowing how to track is a useful skill for:

- Finding a missing person.
- Knowing where your enemy is so you can avoid him.
- Tracking a thief and/or recovering your goods.
- Leading you to find safety among people after you escape capture.
- Tracking animals to hunt for food in a survival situation.

Tracking effectively involves observing signs of presence and then correctly arranging those signs into a story of where your target went.

Becoming a skilled tracker takes practice. You need knowledge of the environment, and/or of the person or animal you are tracking.

What follows are outlines of (very) basic tracking skills.

Signs of Presence

A sign of presence is any disturbance of the natural surroundings. Look for specific signs related to who (or what) you are tracking, such as footprints that are a particular shape or size, or are in a given pattern.

Examples of what to look for include:

- The absence of animals.
- Any sign of humans, such as fabric, litter, fire, shelter construction, etc.
- Bodily fluids (blood, urine, poop, mucus, etc.).
- Broken spiderwebs.
- Damaged foliage.
- Discarded food.
- Footprints.

- Overturned rocks/pebbles.
- Scuff marks from someone/something leaning on trees or climbing over stuff.
- Signs that food has been taken, such as, picked fruit.
- The transfer of soil from one place to another.
- Upturned soil.
- Vegetation pushed down into an unnatural position.

Track Traps

Track traps are places where signs of presence, such as the transfer of water onto rock, are easier to spot ,. Examples of track traps include mud, snow, sand, soft dirt, and fluid.

Look for signs of presence in track traps first. If you can't find any, move onto harder terrain.

Basic Tracking Method

Find an initial track and document it. Draw a sketch and note the length, width, tread patterns, etc.

Find the next track, which is probably about a step-length away. Ensure it is the same as the first track by referring to your notes.

Additional trackers can look ahead for matching tracks while the original tracker continues tracking step by step, as outlined above. When they find one, the original tracker can mark his latest found track and move up to confirm the new track. If it's a match, he can continue tracking from the new point while the additional trackers look ahead again.

Lost Track

If you lose a track, go back to the last positive sign and mark it with something, like a bright ribbon.

Scan the foreground all around you for the next track.

If you can't find it in your initial scan, walk in the most likely direction of travel to see if you can pick it up.

If you do not find it within 100m (330ft), go back to the last positive sign (which you marked) and try a 360-degree sweep. Make ever-increasing circles outward until you pick up the next track.

Determining Direction

Here are some ways to determine which way your target is headed:

- Animals run away from close danger (e.g., humans).
- Foliage bends in the direction of travel.
- Fluid splatters in the direction of travel (e.g., blood).
- Soil scatters in the direction of travel.

These signs are more reliable than obvious footprints if a person has reason to fool you by walking backwards.

Determining Group Size

When tracking an unknown number of people, use the following method to determine the group size. It requires that you track prints.

- Draw a line behind one print.
- Draw a second line 1.5m (5ft) in front of the first line. Make it 1m (3ft) if you're looking for children.
- Count all full and partial prints between the two lines. Round up if you have an odd number.
- Halve the number.

This will give you a rough estimate of how many people are in the group.

Additional Tracking Tips

- Footprints that are far apart and deeper in the toe or the heel indicate running. Those that are and closer together indicate walking.
- Impressions that are close but deep indicate a person is carrying something.
- If one foot leaves a deeper impression than the other, he may be injured.
- The fresher the track, the closer your target is. The top edges can dry within minutes, but actual erosion takes at least 12 hours.
- When looking for signs ahead, look 15m (50ft) in front of you.
- Higher positions can reveal other signs of presence. Climb a tree to look for them.
- Use your other senses (smell and hearing) too.
- Never walk over tracks. It will confuse you if you need to go back to them.
- Pay closer attention near water sources.
- As you collect evidence, put together a story of your target's condition and where he's going.
- Be aware of false signs, traps, and ambush.
- Signs of your target disguising his tracks may be an indication of a rest spot, change of direction, or ambush.
- Low angles of light make tracks easier to spot. This means the best times to track are in the early morning and late afternoon. Put yourself between the track and the sun, and get low so you can see the shadows.
- Make sure you don't get lost.

Related Chapters:

- Search and Rescue

ESCAPING CAPTURE

PRELIMINARIES

Once you are abducted, your best chance for survival is if you escape or are rescued within the first 24 hours.

When your initial effort of fighting off your kidnappers has failed, act submissive. Look down and do as you are told (within reason) so they don't restrain you more than they already have. Lull them into complacency, and then escape as soon as the right opportunity presents itself.

Note: If you expect to be tortured and killed immediately upon capture, you may as well fight to the death.

The earlier you escape the better, because:

- The longer you remain in captivity, the more thoroughly you will be searched.
- The longer you remain a prisoner, the greater chance you will be sent to a more secure area.

However, you must choose your escape opportunity wisely. If you're caught, you'll be punished and security will increase.

STALLING

There are several tactics you can use to stall for time until help arrives and/or to create chances for escape.

When you are in a standoff scenario and you know your capture is inevitable, try to negotiate your surrender. Even if you do not expect help to arrive, you can try to give yourself better terms of imprisonment.

Another option is to feign injury and ask for medical treatment. Pretending to have a seizure or acting insane is often enough to make any type of criminal leave you alone if you are a random target.

A final stalling tactic is to use a "restricted access" ploy. This is good with criminals seeking material gain. Tell them you have a safety deposit box with valuables that only you can access. When they take you there, use the opportunity to escape.

Related Chapters:

- Bargaining

GATHERING INFORMATION

As soon as you're taken, start using all your senses to find out as much about your captors and where you are going as you can. Take note of their language, the number of people, the style of clothing, their names, organization, motivation, equipment, personalities, etc.

When in a vehicle, try to determine your traveling speed, surrounding noises, the time in vehicle, turns, direction, etc.

Once in captivity, look out for exits, security (and lack thereof), location, weather, the surrounding environment, useful resources, other captives, your captors' routines, etc.

Related Chapters:

- Traveling

LEAVING CLUES

Once in captivity, your best chance of escape is to get rescued. Make it easier for rescuers to track you by leaving clues of your presence in every vehicle and room in which you're held. For example, you can:

- Construct or draw arrows.
- Leave DNA.
- Drop notes.
- Leave behind pieces of clothing.
- Build rock piles.

Leaving and/or collecting DNA for investigators will help them find you, and is also useful in convicting your captor(s) later.

Any bodily fluids leave traces of DNA (blood, vomit, urine, spit, etc.) ,as does hair. When leaving your own DNA, put it in places that your captor will (hopefully) not clean, such as under/in/behind furniture, in door hinges, in air vents, on walls and in corners. Tell your family about these tactics so they can advise the police to look for your DNA in unusual places.

When collecting your captor's DNA, you want to "store" it on you so it won't get washed off. Scratching him hard enough to draw blood will put it under your fingernails, and it is likely to stay there unless scrubbed off.

Another thing you can do is wipe his sweat (or other body fluids) under your body hair. When you shower, avoid washing those areas, unless you have been captured for a long time, in which case you need to maintain hygiene for your health.

ENDURING CAPTIVITY

When an early escape is not possible, you must concentrate on surviving captivity until you can escape or are rescued.

A lot of the information in this section also applies to surviving a hostage situation or government detention facility, such as a POW camp or prison.

Acceptance

Accept the fact you're a prisoner. Move on from the self-pity and anger so you can focus on survival and escape.

Be the Gray Man

When you're first captured, and especially when you're in a group, do nothing to attract extra attention to yourself. Keep calm, quiet, emotionless, and compliant. Stay still, with your eyes on the floor.

The Will to Live

A big part of survival is maintaining your will to live and a strong belief that you will survive. Remember your reasons for living (e.g., loved ones) and have faith in yourself, your abilities, and your god if you have one. No matter what happens, do not give up your will to live, and always be prepared to seize the moment you can escape, even if it takes years.

Although escape becomes harder the longer you wait, the longer you are held captive, the more likely it is that you will eventually get out alive. If your captors' intent is to kill you, they will do it sooner rather than later.

Stay Healthy and Mentally Active

Creating good mental and physical habits helps you sustain a will to live. It also keeps you mentally sharp and physically fit so you seize opportunities to escape.

A productive way to exercise your mind is by planning your escape. Use all your senses to collect information and probe your captors to discover who you can exploit. Other than constant scheming, make use of any entertainment you can get hold of, such as e.g., reading.

Establish a physical routine and do it regularly. Do whatever you have room for. Pushups, sit ups, and stretching are great exercises that don't require much space.

Eat anything you are given, as long as it isn't poisonous. Refusing food as a protest is not a good strategy for long-term survival, and being a gracious prisoner may win you extra favors.

Humanize Yourself

The more human you are, the harder it is to hurt you. Naming yourself is a good start. It's harder to kill or beat something with a name. No matter what your captors do, remain calm and polite. An overly emotional or difficult prisoner is easier to treat badly, so maintain your dignity. Don't beg, cry, soil yourself, etc.

Make Friends with Your Captors

Social contact has psychological benefit and furthers your humanization.

Developing bonds can also help your escape. It's easier to extract information from someone you have rapport with. Target those who seem more sympathetic towards you.

You're also more likely to get extra comforts if you're friendly. Start by asking for small things, like a drink or blanket, then get more ambitious, requesting extra food or entertainment. Don't push it, or you may get cut off completely.

Making friends with your captors is useful, but it's important to remember that they're still the enemy. Do not hesitate to hurt any of them during an escape attempt.

Work with Other Prisoners

Making friends with other prisoners has several advantages. It is psychologically beneficial, you can work together to escape, bargain on each other's behalf (if one of you is getting punished), and send for help if one of you escapes.

However, you must choose your friends wisely. Not everyone will act for the good of the group, especially petty criminals.

In a group situation—where there are multiple hostages, or in a POW camp or prison—it is better to keep an "us vs them" mentality. Taking sides with the guards may get you killed by other prisoners. Do not do anything (including accepting favorable treatment) that will be harmful to other prisoners. This includes divulging information.

When it comes to group control, either take command or obey and back up those in charge (not the enemy).

When you're in prison with criminals, it becomes a game of wits against guards and other prisoners. Choose your friends wisely, and don't fall into pettiness and manipulation.

Interrogation

When subject to interrogation, give up as little useful information as possible while remaining calm and polite. Do this by speaking only when told to and giving short answers.

Avoid eye contact with your interrogator. If you are forced to look, stare at his forehead.

Be wary of any deals you are offered.

Unless you'll be tortured or killed if you refuse, avoid:

- Confessing in any way.
- Making propaganda broadcasts.
- Speaking out against your cause (verbally or in writing) if you are a political prisoner.

When you're held by a government or professional political organization, treat all conversations with your captors as interrogations, even when they seem casual.

Contact the Outside World

Do everything you can to contact the outside world. Appeal to family, friends, lawyers, and other sympathizers, so they can start making plans to get you free. Have them constantly inquire about your health and wellbeing. Allow your captors to take a picture of your face so the authorities can ID you.

Related Chapters:

- Planning Your Escape
- Active Listening

PLANNING YOUR ESCAPE

Start planning your escape from the start and never stop, no matter how long you are held captive.

Apart from everything explained in the Planning and Preparation chapter, there are two main things to consider for your escape: when to go and what route you'll take.

Planning the "perfect" route and time for escape is good, but don't hesitate to take chance opportunities when they arise.

If you fail an escape attempt, expect to be beaten. Feign injury and/or exhaustion so you appear to be less of a threat.

When to Escape

When you're abducted for ransom, you're likely to be released as soon as your captors' demands are met. Attempting a risky escape may not be worth it, especially if you're being held in a remote location where you will need to survive the elements to get to safety.

If you're the captive of a sexual predator, escape ASAP. Otherwise, probably be killed once you've served your purpose, or will live a life of misery.

When your captors suddenly do any of the following, your time may be limited:

- Stop feeding you.
- Treat you more harshly.
- Get desperate or frightened.

In this case, attempt escape even if your chances are not good.

Anytime you are moved out of your cell is an opportunity for gathering intel, preparing an escape, or actually escaping, especially if the move is a routine occurrence.

Good times to escape include:

- When they won't check on you.
- Night time.
- During bad weather.

Choosing Your Route

When choosing a route, go primarily for stealth. Keep to places where you are less likely to be seen, and where there are few warning systems such as alarms, booby traps, lights, or dogs. Consider what distractions you can make and what obstacles you can put in your enemies' path.

If possible, plan alternate routes as well. Have one directly opposite and one 90 degrees from your primary route.

SURVIVE A RESCUE ATTEMPT OR RELEASE

The authorities may send in a tactical team to rescue you. This is great, if you survive the rescue.

If you have time,, get to a safer part of the room as soon as you are aware of the rescue attempt. Choose somewhere:

- Under or behind cover.
- Away from doors and windows.

Then go into the grenade survival position:

- Lie on your stomach.
- Point your feet towards the likely entry or blast point.
- Cross your legs and cover your ears.
- Keep your elbows tight against your rib cage.
- Open your mouth a little.

Once any explosions or bullets have passed, get on your back and spread your hands and legs out to show they are empty.

To prevent yourself from being mistaken for a bad guy, do not:

- Stand up.
- Run away from the rescuers.

- Pick up a weapon.
- Try to help the rescuers.

Be prepared for hostile treatment by rescue forces until you are positively identified.

Getting Released

If your captors are releasing you for whatever reason, follow their instructions.

CARS

Chances are that you'll come across a car while being taken and/or during your escape. In this section, you'll learn a variety of car-related tactics such as general safety, escaping a car, evasive driving techniques, and more.

ESCAPING CARS

Your chances of escape dwindle the further you go away from where you were originally abducted.

If you couldn't fight off your attacker initially, try your best to escape the car. Once he gets you to a secure place, it will be much harder.

Escaping a Car Trunk

There are a few things you can try to escape when shoved into the trunk of a car:

- Pull the emergency trunk release lever. These are common in newer cars.
- In older cars, pull the release cable.
- Press your back against the trunk roof and use your arms and legs to push it open.
- Use the car jack to force it open.
- Disconnect the brake light and kick it out. Put your hand through the hole to signal for help.
- Kick through the back seat.

When you are getting put into the trunk, try to position yourself so you can access your escape tools.

Jumping from a Moving Car

Jumping from a moving car is dangerous, but is better than abduction. Before attempting to jump out, ensure the door is unlocked.

Prepare to jump out at the safest time:

- Any more than 30mph (50kmh) is too fast. Choose a time

when you are stopped, beginning to accelerate, or just before turning a corner.
- Ensure there is nothing in your jump path. You will continue to move in the same direction and at the same speed the car is moving.
- Landing on a soft surface, such as grass, is preferable.

If possible, pad your clothes with something soft, like newspapers.

When it's time to jump, open the door fully so it's less likely to close on you. Leap as far out as possible on an angle, in the opposite direction from the one in which the car is moving.

If the car is turning, jump from the side opposite from the one to which it is turning. That means if you're sitting on the right, wait for a left-hand turn.

Roll up into a ball and tuck your chin to protect your head. Try to land on your back and roll when you land.

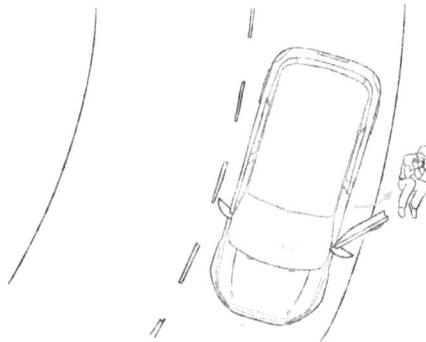

Disabled Driver

To take control of a car when your driver is incapacitated, use your leg to push his leg out of the way. Gain control of the gas pedal and steer the car to safety.

Kick Out a Car Window

Car windows are tough, especially the front and back windshields, so do not try to kick out one of those.

To kick out a side window, lie on your back with your feet facing the window.

Use both feet together to kick at the lower-right part. Your feet will bounce off if you try to kick it in the center.

Escape a Sinking Car

If you find yourself in a car heading for water, brace for impact.

As soon as the initial collision with the water is over, open your window. You want to do this as soon as possible—even before hitting the water, if possible.

Try to climb out before the car starts to sink.

When the window won't open enough, smash it with a glass-breaking tool or heavy object such as a steering-wheel lock, or kick it as previously described.

If the car starts to sink before you can escape, wait until water stops rushing in and swim out the window.

As a last resort, wait until the car fills up with water. This will release the pressure, and you will be able to open the door.

When you need to hold your breath, completely empty your lungs first then take a deep breath so your body is full of fresh air. Try to stay calm so you can hold your breath longer.

Related Chapters:

- Covert Survival Kit
- Car Jack

DISABLE YOUR ENEMY'S CAR

There are several ways to disable your enemy's car so he can't chase you. Do as many of the following things as possible given the time and access you have.

In general, cut wires, drain fluids, and rip stuff out of the engine. Here are some more specific suggestions:

- Puncture the tires.
- Remove the tire bolts.
- Stuff something in the exhaust. Pack it tight.
- Stab something sharp through the radiator.
- Remove the spark plug leads.
- Block the air intake filter with cloth.
- Flood the air intake filter with a hose.
- Remove the battery.
- Remove the starter.
- Light a rag and leave it in the gas tank.
- Contaminate the gas tank with, dish soap, sugar, water, or dirt.

STEALING CARS

Just because there is a car to take doesn't always mean you should. You'll cover a lot more distance in one, but it will also be easier to track you.

If you do decide to take a car and have a choice, the best one to steal is one that:

- Is easy to steal.
- Doesn't stand out. It's too dirty or clean, and it has few obvious identification marks (bumper stickers, dents, bright paint, etc.).
- Is lower to the ground. Taller cars are easier to flip in a chase.

Obtaining Keys

Getting the keys is the best way to get a car. There are various ways you can do this.

- Steal the keys—by pickpocketing their owner or by taking them from a valet booth, for example.
- Carjack someone. This is good for a quick getaway, because the car is already running. Gas stations and ATMs are good places to do this, as the driver may leave his keys in the car while he gets out.
- Use master keys. You can steal them from tow trucks. Sometimes keys from a different car but the same manufacturer will work.
- Find spare or valet keys in an unlocked car. Check the center console, glove box, under the mats, or on the visor. Valet keys are often in the owner's manual.

Gaining Entry

When the car is not unlocked and you do not have the keys, you will need to break into the car before you can take it.

To smash a window, use something hard to hit it at the corners. Side windows are the weakest. If you have the resources, tape a large X over the window before smashing it. This helps to contain the sound and prevents shattering.

Alternatively, you can try to pick the lock. For a pull-up-style lock, use a shoelace. First, tie a small loop in it using a slip knot. Shimmy the shoelace into the door via the upper corner of the window frame. Once it's inside the car, maneuver it above the lock.

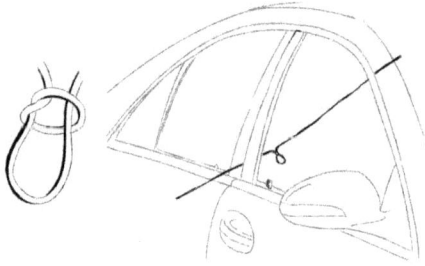

Tighten the loop around the lock and pull up on both ends.

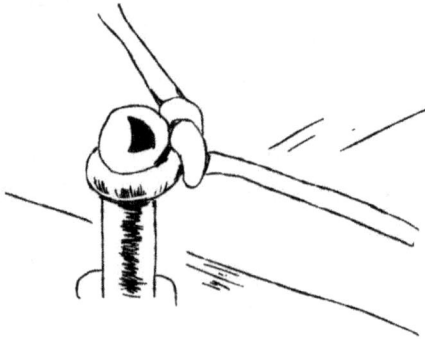

A final method that doesn't need any special equipment is to use a coat-hanger and a shoe.

Pry the upper corner of the door open and wedge your shoe in the gap. Use a straightened-out coat-hanger to unlock the door.

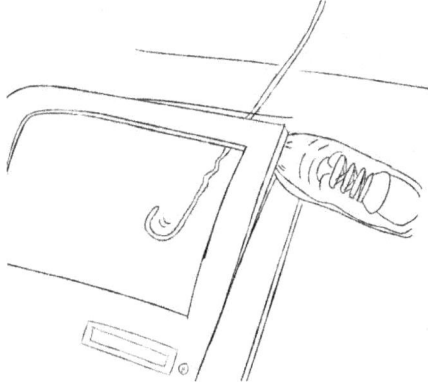

Starting a Car

Newer cars are almost impossible to start without the key or special equipment.

If a car was made before 1999, you may be able to pop or hotwire it. The older the car, the better your chances of success.

Don't practice this on a car you need to use. It will mess it up.

Before starting, put the car in neutral and with the parking brake on. If it's an automatic, put it in park.

Lock-Popping

To lock-pop a car, you need:

- A flat-head screwdriver.
- A hammer.
- Pliers (optional).

Insert the screwdriver into ignition and force it in with the hammer. Use the pliers to turn it.

Hotwiring

To hotwire a car, you'll need:

- Wire cutters and strippers.
- Pliers.
- Flat-head and Phillips-head screwdrivers.
- A hammer.
- Insulated gloves.
- Electrical tape.

Remove the plastic panel above and below the steering column. You can either unscrew it or smash it off.

Select the bundle of wires going into the steering column. It will be five wires connecting to the ignition cylinder (where you insert the key).

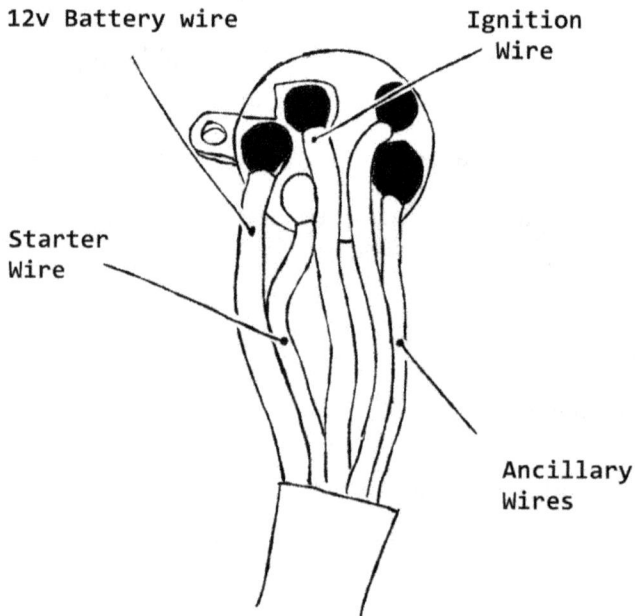

12v Battery wire

Ignition Wire

Starter Wire

Ancillary Wires

Pull out the ignition cylinder and cut the first three wires in the sequence (battery, starter, and ignition) loose. The colors will vary depending on the car.

Expose the wires, but do not touch them with your bare hands. Use insulated gloves or a cloth.

Join the battery and ignition wires together to light the dashboard up. Touch the starter wire to the battery/ignition wire to start the car.

If the car has two starter wires, touch those two together—and not to the battery/ignition wire.

Once the car starts, wrap the starter wire(s) with electrical tape. This prevents it from touching you or the other wires. To turn the engine off, separate the battery and ignition wires.

Steering Lock

If a car has a steering lock, turn the wheel forcefully in one direction until the lock-pins break. Alternatively, locate a gap at the center of the steering column, between the wheel and the column itself. Force your flat-head screwdriver into the gap to push the locking pin away.

Related Chapters:

- Picking Pockets
- Picking Locks

CAR SAFETY

Here are some tips for general car safety.

Service your personal car regularly and check the basics (oil, water, tire pressure, tight lug nuts, etc.) weekly. The optimal tire pressure for normal driving is 10 percent under the recommended PSI according to the tire, and not the car manual.

Always keep your gas tank at least 1/4 full, and tape a razor blade to the shoulder strap as an escape tool. A glass-breaker is another life-saving tool to keep within reaching distance.

Adjust your side mirrors for maximum peripheral vision. If you can see any part of the back of your car, you need to push them out.

When driving, always wear a seat belt. Hold the steering wheel in an overhand grip (thumb next to your fingers) at the 9 and 3 o'clock points. Never cross your hands or steer with your palm. Use shuffle steering.

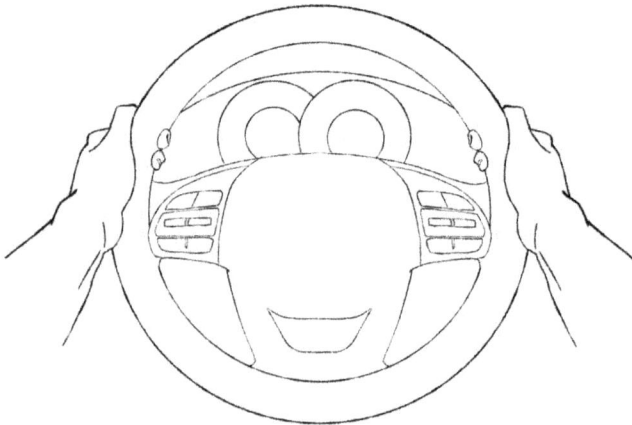

To avoid becoming a victim of road rage, follow the road rules. If you make a mistake, smile and mouth "sorry" at the driver you affected as you drive away.

If you get a flat tire, turn on your hazard lights and drive slowly on the shoulder of the road until you get somewhere safe enough to change it. The side of the road is not safe!

Learning basic car repair may save your life. At a minimum, you should know how to:

- Change a tire.
- Pump the tires.
- Connect the battery.
- Jump-start a dead battery.
- Check and top up the car fluids.

If you break down in an isolated or otherwise unsafe area, lock yourself in the car and call for help.

For passengers, the safest place in a car is behind the driver's seat. In the event of a crash:

- Tighten your seat belt as much as possible.
- Fold your arms across your body.
- Sit upright, with your back and head pushed back into your seat.
- Relax your body.

Car Safety Kit

Keep the following items in your car in case of an emergency.

- Small fire extinguisher.
- Spare tire.
- Tire jack.
- Weapons—one in the trunk, and one within reach of the driver's seat.
- Jumper cables.
- Ropes for towing.
- Emergency fluids (oil, gas, and coolant).

- Three days of food and water.
- First aid kit.
- Cold-weather gear (blankets, extra clothing, poncho).

Car Modifications

Making some car modifications increases performance, reliability, and safety. Here are the minimum additions you should make:

- Car alarm and immobilizer.
- GPS navigation.
- GPS tracking.
- Radial tires filled with run-flat foam.
- Quartz-iodine lights.
- Stainless steel brake lines.
- Locking gas cap.
- Thick bolt through the tailpipe.
- Wide-angled electric side mirrors.

If you will be driving in rough terrain, upgrade the following to heavy-duty versions:

- Radiator.
- Shocks and springs.
- Steering pump.
- Battery.

Related Chapters:

- Car Accidents

EVASIVE DRIVING

There are a variety of evasive driving techniques in this chapter. Some are useful for everyday life, but others can be dangerous. Practice them safely.

Only try these techniques in cars with a lower center of gravity. SUVs and minivans are likely to flip.

Before you endanger yourself and other drivers, ensure you have a tail.

Two-Foot Driving

If you are driving a car with automatic transmission, using two-foot driving will increase your reaction time.

To do it, use one foot for the brakes and the other to accelerate, as opposed to using one foot for both. Use the balls of your feet to depress the pedals.

Threshold Braking

Threshold braking is a technique for slowing down faster. It will improve your cornering and other precise maneuvers.

Apply gradual but firm pressure on the brake until just before wheels lock or ABS kicks in. When/if the wheels lock, release the brake a little, then reapply it with slightly less pressure. If your tires screech, you need to release the brake—but don't wait for that.

Outrunning a Pursuer

Unless you know your car will outrun a pursuer's and you're on open roads, keep your speed under 100km/h (65mph). If you're going any faster, and you're likely to crash.

Prevent your pursuer from pulling alongside you by blocking his way. If he gets next to you, he'll have a clear shot or be likely to ram you.

If he's shooting at you, slalom (weave) to avoid the bullets. When shooting back, aim for the driver and/or his front tires. It's best if a passenger does the shooting. Put him in the back seat so he can shoot in any direction.

To jump a curb, slow down to under 70km/h (45mph) and approach it at a 45-degree angle.

As a last resort, go off-road. Drive extra-carefully, as there will be many additional obstacles (dips, rocks, etc.). When you can't go any further, get out and take cover so you can ambush your pursuers.

Cornering

Good cornering is all about when you take the apex. The apex is the point where your wheels are closest to the inside edge of the corner.

When you have several car lengths between you and your pursuer, aim for a late apex. This means you will slow down before entering the corner, but will exit it faster, and the faster you exit, the faster you'll on the straight after it.

If your pursuer is within several car lengths it is better to take an early apex. Otherwise, he may catch you in the turn when you slow down.

Here are some specific cornering techniques. They assume you want to take a late apex.

To take a 90-degree turn, start as far to the outside as possible. Use threshold braking as you approach, and release your brakes when you're into the first third of your turn. Speed up as you exit the turn.

In s-bends, drive in as close to a straight line as you can.

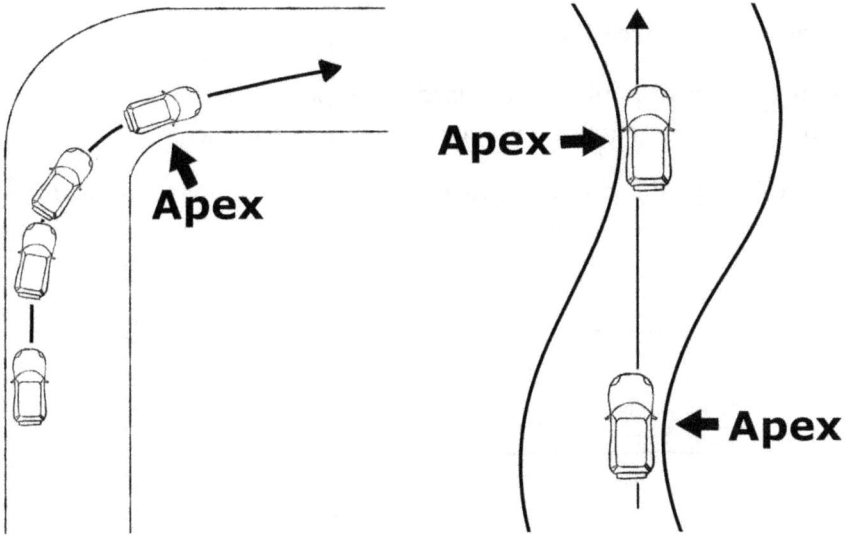

To get around a hairpin turn, start wide on the first half, and treat the second half like a 90-degree turn. Take it slower than other turns.

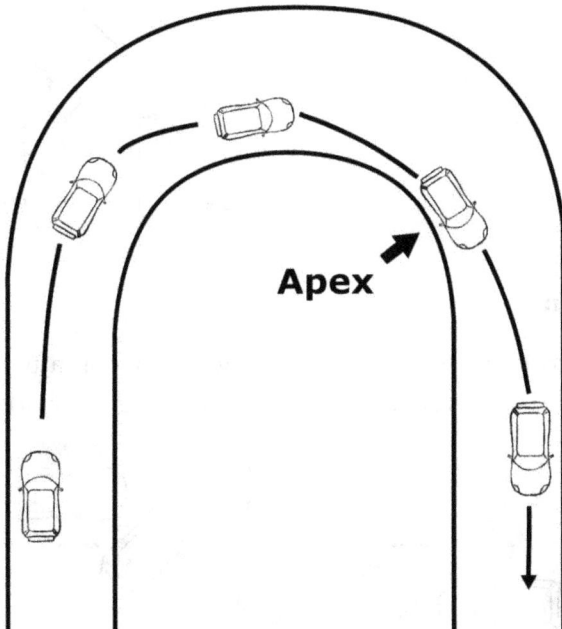

Bootlegger's Three-Point Turn

This is a variation of the standard three-point turn. It allows you to reverse your direction after a curve on a narrow road.

Immediately after the bend, turn into an intersecting road (or driveway). Once your pursuer passes you, reverse and drive off in the opposite direction.

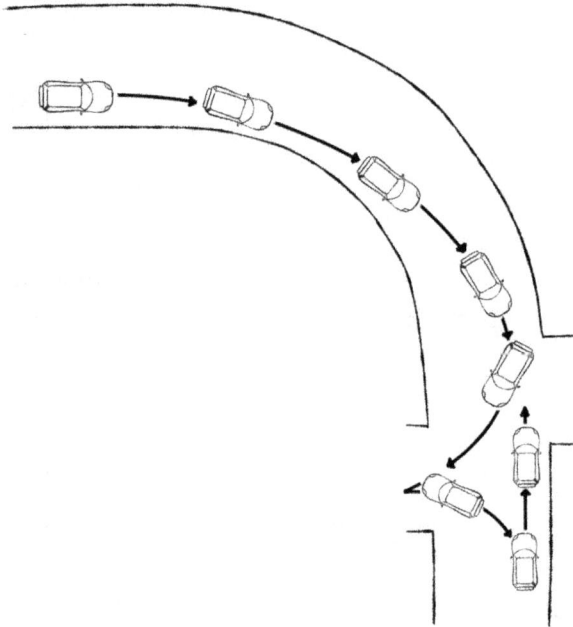

Bootleg Turn

The standard bootleg turn is a 180-degree turn on a two-lane road. It's good to use after a blind corner or on a long, undivided bridge.

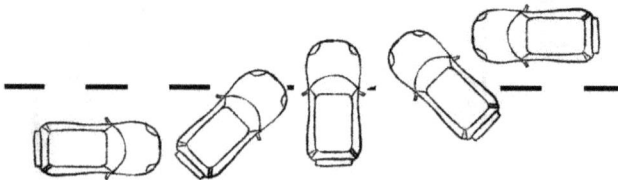

When practicing this maneuver, inflate your tires to 10psi over the recommended maximum. This will prevent them from blowing. Expect your front tires to wear out quickly.

To prevent flipping or losing control, do not go over 50km/h (30mph).

If you want to turn left (as in the image) do the following in quick succession:

- Place one hand on top of the steering wheel and the other on the emergency brake (or hand brake). It is important to use the hand/emergency brake. The normal foot brake will lock up the front tires.
- Turn the wheel a little to the right.
- Apply brake and simultaneously turn the wheel sharp to the left until your hand is near the 6 o'clock position.

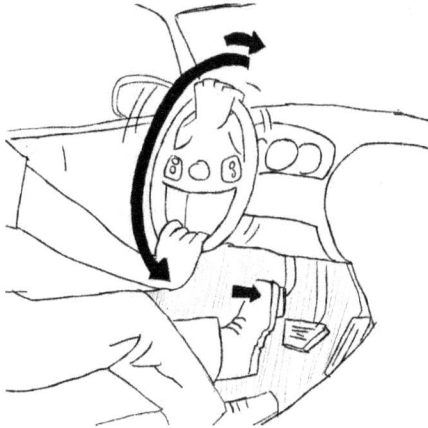

If in a manual car, apply the clutch as you put on the brake.

When the car is at 90 degrees, release the emergency brake, straighten the wheel, shift into low gear (manual transmission), and accelerate.

Do not floor the accelerator.

Reverse 180

To make a 180-degree turn on a two-lane road while going backwards, use the reverse 180.

This is good against roadblocks, and with enough practice you can do it in one lane.

As with the bootleg turn, inflate your tires to 10psi over the recommended maximum and don't go more than 50km/h.

If you want to turn left (that is, if the road space is to your left):

- Place your hand at 4 o'clock (7 o'clock if you want to turn right) and put your other hand on the gear shift.
- Accelerate in reverse to about 40km/h (25mph). Use your mirrors to look behind you, as opposed to turning your head.
- Turn the wheel a little to the right, then take your foot off the accelerator, shift to neutral, and turn the wheel sharply to the left as far as you can. Do not use the brakes.
- When the car is at 90 degrees, shift into a low/forward gear, straighten the wheel, and accelerate.

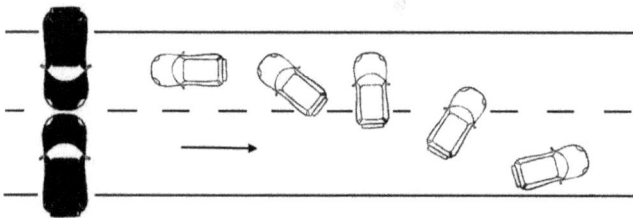

The Cut

The cut is a maneuver you can use to lose a pursuer in traffic. Without indicating, turn in front of oncoming traffic in the opposite lane.

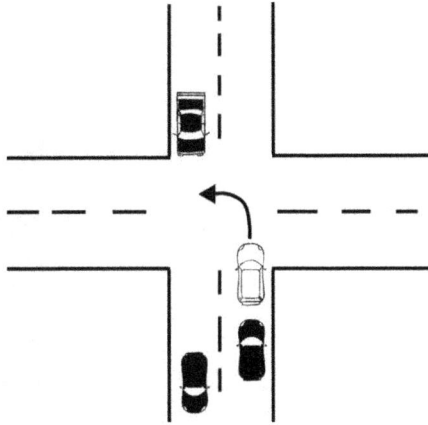

Ram Through a Car Block

When you encounter a roadblock made of cars, it's preferable to go around it.

If that's not possible, your aim is to push through it.

As you approach, slow down to under 30km/h (20mph). This will let you avoid disabling your car on impact and give any guards the impression you're stopping.

Aim to hit the corner of your car on the corner of the barricading car. Any corner-to-corner contact will work, so consider what's behind the barricading car. If there's nothing else to consider, then pressing your passenger side (furthest from you) to the other car's rear corner (the lightest side) is ideal.

Keep your foot on the gas at steady pressure until you are through, then accelerate.

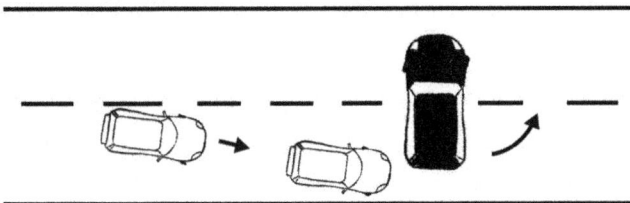

When there are two cars, aim either for the car that's the easiest to get past or the center of the gap.

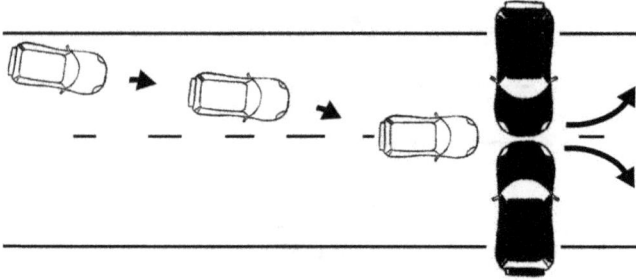

Take Out Another Car

If you manage to get behind your pursuer, you can use the following techniques to ram him off the road. Unless he's going under 45km/h (30mph) he'll probably crash.

The first method is the precision immobilization technique (PIT). Get your front bumper in line with his rear wheel. Maintain your speed and nudge your front bumper into his rear wheel. Immediately apply your brakes and steer around him as he spins in front of you.

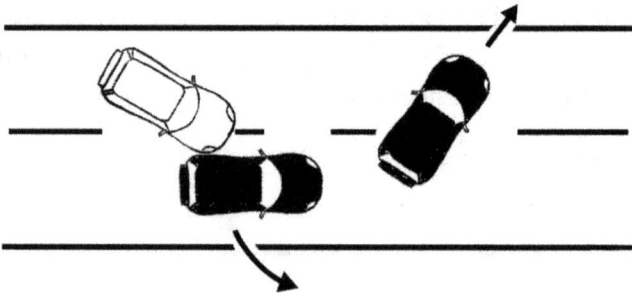

To do the second method, start directly behind him and speed up, so you are going about 20 km/h (10mph) faster than he is. Hit the corner of your front bumper into the opposite side of his rear bumper. It is a hit, not a push.

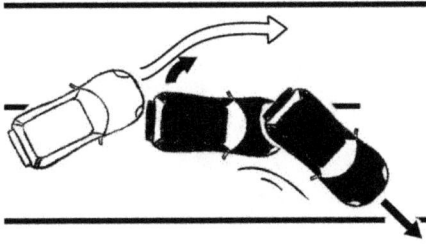

For the final method, accelerate to overtake him. As you pass him, nudge the middle of your car into the corner of his front bumper.

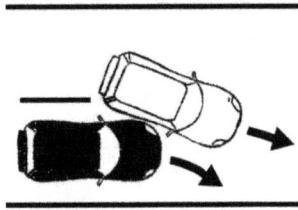

Related Chapters:

- Stalkers

NEGOTIATION

Knowing some basic negotiation tactics is useful in many areas of life, from consumer purchases to business to convincing your kids to go to sleep. In the context of the subjects discussed in this book, it can help you get extra benefits as a prisoner, or bargain for your release or that of a loved one.

The basic idea of any negotiation is to discover what the other party wants, figure out how to give it to them, and exchange it for what you want. Negotiation is a non-linear process. To do it successfully, you need to acquire five main skills:

- Active listening.
- Eliciting information.
- Detecting lies.
- Overcoming barriers.
- Bargaining.

Each of these skills is useful in its own right. A skilled negotiator will use them to complement each other.

But before you can use any of them, you need to set your minimum goal. Make it something realistic that you'll be happy to get. During negotiations, always aim to get the best deal possible, but never go below your goal. If you're not getting the offer you want, walk away.

ACTIVE LISTENING

Active listening is the basis for creating any positive relationship. It builds rapport and elicits information at the same time.

To practice active listening, devote all your attention to understanding what the person is saying, both verbally and with his tone of voice and body language.

In a formal negotiation, pay special attention at the start and end of the meeting, as well as at times of disruption. These are the times you can observe him unguarded.

When he stops speaking, show understanding by echoing what he is saying back to him. Do this by repeating the last three, or the critical one to three words he said, prefaced by the phrase "You think/want/feel ... " Another way to echo is to infer and repeat his feelings back to him, prefaced by the words "It seems/sound/looks like you ..." You can use echoing as a statement or a question. The only difference is your inflection.

Always pause for at least five seconds after echoing. This will give him time to process, and in most cases, he will fill the silence. Do not interrupt him to echo again or for anything else. When he is talking at length, use simple phrases such as "Yes," "OK," and "I see" to show that you're paying attention.

In the beginning, you may need to ask questions to get him going. Start with common subjects such as family and interests. Let him escalate to his goals, values, and wants. You can encourage this a little to speed the process up.

Once you know what he wants (power, money, sex, etc.), figure out how you can provide or withhold it, and use it as leverage when negotiating. You can leverage his values as well . No one wants to be a hypocrite.

Show genuine interest in his goals and in his ability to achieve them. This is a great rapport-builder.

When your opponent says "That's right," it means you have correctly echoed what he feels. It is more genuine and engaged than "Yes," which people often use to humor others.

The ultimate "That's right" is when you successfully summarize his overall view point. It is like combining and paraphrasing all your echo statements together in a way that encapsulates how he views the situation.

Note: "That's right" is different from "You're right." "You're right" is akin to "Yes."

Building Rapport

It is much easier to broker a deal with someone you are friendly with. Aim to build rapport from the start and continue to do it throughout the entire interaction.

Other than active listening, there are some other general things you can do to build rapport.

Make a positive first impression when you first meet someone by looking them in the eye and saying "Hi, (name)" while smiling genuinely. This is irrelevant with a hostile person (such as your kidnapper) but it is good to know for everyday interactions.

Be polite and respectful. "Please" and "thank" you go a long way. Being critical, argumentative, or giving unsolicited advice is not polite. Encouragement and compliments are, but only if they are genuine. No one likes a suck-up. You do not have to agree with everything the other person says, but do not be rude.

Be responsible and trustworthy. Admit when you make a mistake (unless there will be legal ramifications) and follow through on what you say. This means you must be honest about what you can and can't do. Smile as you speak (even on the phone) to project a positive attitude.

Avoid the word "I." Constantly talking about yourself or the things that you want will turn him off you and the deal.

ELICIT INFORMATION

Even when not negotiating, you want to get as much information from your captors (or others) as you can. You never know what you might discover that can help you escape.

When targeting someone for information, go for the weak link if possible. This is usually whoever is nicest to you (gives you extra food, for example). Start by using active listening. When that's not enough, try some of these tactics.

Be on the lookout for people using these tactics against you, too, whether in captivity (interrogation) or in everyday life.

Get Help

Often, people are happy to show you how to do something without realizing they shouldn't be teaching you.

Flatter a Target

Flatter your target about what he did (or what you think he did), and he may be happy to tell you exactly how he did it.

Correct Me

Make a false statement to prompt the correct answer.

Tell Me More

When he touches on a subject of interest, encourage him to speak more about it with an open-ended question, such as "Oh, that's not good. Why did that happen?"

Sharing Knowledge

Show that you have knowledge of a subject, and he may help you fill in the gaps or tell you what he knows just to be part of the conversation.

Indirect Questions

People may be defensive when asked things directly. Ask indirect questions to get the answer. For example, instead of "What did Ryan do wrong?" ask "What would you have done differently?"

Hurt Feelings

People may withhold information or tell white lies to protect your feelings. Reassure them your feelings won't be hurt and ask for the brutal truth.

Guess

When someone says "I don't know," ask something like, "What's your best guess?"

Confide

Confess some similar wrongdoing to a target to build trust. He may confess his wrong doing in return.

What Happened Isn't Important

If you suspect someone is lying or did something wrong, tell them that you don't care about the act, but rather that honesty in your relationship or his motivation for doing it (e.g., if it was an accident) is more important.

Give a Reason

Sometimes people need a little push to divulge information. Use a "because statement," like "I need to know if ... because ..." Make the reason a good/serious one.

Last Chance

Inform your target that if he doesn't tell you now, he won't get another chance. Give him a reason why there won't be another chance, or explain what could happen if he doesn't talk.

Attack Your Target's Ego

Infer that your target probably doesn't know the answer. He may give it to you as proof that he does.

Help Your Target Out

Tell your target that you can help him get out of a bad situation, but you need to know the facts first.

Related Chapters:

- Active Listening

DETECT LIES

While eliciting information, you will need to know what is true or not. These skills are also useful for detecting lies in general.

There are some common behavioral signs that people may display when they're lying:

- Telling a mixed-up story with contradictions.
- Answering your question with a question or some other non-answer.
- Blaming others.
- Blocking further investigation.
- Not being able to provide proof.
- Confirming, denying, or not correcting false facts.
- Constantly referring to other people with third-person pronouns (he, she, they, etc.).
- Fumbling over words as their brain is formulating the lie.
- Pointing away from you and/or jiggling them. This is an indication the person wants to get away.
- Being more interested in the consequences than the story.
- Knowing things he shouldn't.
- Moving less or freezing completely.
- Overreacting when confronted.
- Presenting themselves as trustworthy instead of answering questions directly (telling you about their good deeds or religious nature, for example).
- Staring at you too hard.
- Telling stories that don't match up with other people's Always question accomplices separately.
- Suggesting a lighter punishment for the "unknown" culprit.
- Using a tone of voice and body language that are incongruent with what they say. For example, they say yes but subtly shake their head from side to side.
- Using an excessive number of words.

The above signs can indicate a liar, but they are not very reliable. Even if a person displays several of them, they may be telling the truth.

A more accurate method is to study what a person's behavioral pattern is when he is not lying. To do this, you need to first establish his non-lying behavior as a baseline.

When his mannerisms conflict with this baseline, you can make a judgement if he is lying or not.

Establishing Baseline Behavior

Make your target comfortable, physically and mentally.

Ask simple, open-ended questions he has no reason to lie about.

Study his behavior and take mental note of his mannerisms as he talks.

For example, track whether he's tapping his fingers, looking away from you, biting his nails, or making particular facial expressions. These are his normal mannerisms—assuming he is speaking truthfully.

Now you can ask questions that he may lie in response to. Look for the common signs of a liar, discounting any that you've noticed as part of his normal behavior.

Action

If you think someone is lying, try to get him to tell the lie three times in the same conversation. It is hard to tell the same lie three times in a row, especially if he just made it up.

To do this, echo what he says to you, so he confirms it. You can also ask a question to get him to re-explain that part of his story.

For example, you can ask, "How did you ... again?"

Confronting someone after confirming a lie is not recommended in most cases.

Instead, use the knowledge to make better decisions, and continue to elicit information.

Related Chapters:

- Elicit Information

OVERCOME BARRIERS

A barrier is anything that gets in the way of making a deal. More often than not, both parties in any serious negotiation will have numerous barriers.

It is important to remember that it is the barriers you need to overcome, not the person. No matter what obstacles arise, stay calm and polite and focus on the deal.

Here are some tools you can use to overcome barriers in a negotiation.

Pre-empt Objections

Before entering negotiations, list every objection the other person could have to your offer. To make this easier, imagine yourself in his shoes, with the mindset of wanting to shoot down your offer.

Think of a win-win solution and/or a positive spin to every objection you write down.

Get Him to Say "No"

Get your opposition to say "no" early in the negotiations. This will give him a feeling of control, and once he's said it, he'll be more receptive to negotiations.

If he doesn't say "no" naturally, trigger it by:

- Echoing him incorrectly so he has to disagree with you.
- Framing questions to which the positive answer will be a no, such as "Are you going to hit me?"
- Asking a question that can only be answered negatively, such as "Do you want to get arrested?"

When a person refuses to say no it is an indication that he is indecisive, confused, or that he has a hidden agenda. In this case, it's best to walk away from the negotiation. If you need to revisit it, try to find someone higher up to deal with.

Ask "How?"

"How" questions are the easiest way to uncover solutions to objections, whether they are yours or the other person's. Use them early and often.

Using a "how" question works because it gets him to think up solutions and implementation strategies. This gives him a vested interest, since those were his ideas.

When using a "how" question, do it from a problem-solving frame of mind. Otherwise, it may sound like an accusation.

If a "how" question seems out of place, try a "what" question, like "What can I do to make this problem go away?"

Never ask "why?" It is an accusation.

Use "Because"

People are more likely to comply with your request when you give a reason, and the simplest way to do this is by including the word "because" in your request. For example, you can say, "It may be better to … (action) because … (reason)." Make sure you use a reasonable tone of voice so it comes out as a request and not a demand.

Combining "how" and "because" also works well. For example, say "How do you expect us to …? Because …" Alternatively, replace "because" with "when" (which is essentially the same word in this context), so the question becomes "How can we … when …?"

Be Fair

People are more likely to comply when you give them fair treatment. If you are accused of being unfair, ask "How am I being unfair?" to uncover the objection.

Never accuse your opposition directly of being unfair. It will only make him hostile. Imply it with "how" questions. For example, say, "How am I supposed to ... when you ...?"

Deadlines

People often use deadlines to rush a deal, but they are almost never set in stone.

When threats get specific, start to pay attention. You can judge the specificity of threats by how many of the "four questions" (what, who, when, and how) are answered. The more that are answered, the more specific a threat is.

Declining Offers

Saying "no" directly stifles negotiation and may offend your opposition. There are ways to decline gently, which allows him to make counter-offers without losing face. You can say no without using the word several times before having to take a firm stand. Use them all.

When you first object to something, summarize the situation and use a "how" question, such as "How are we supposed to ...?" or "How do I know if ...?" Do this several times if the situation allows.

If you object to his next offer, mention his generosity, apologize, and decline: "It's a generous offer but I'm sorry, I can't make it work for me."

For your next rejection, apologize and decline: "I'm sorry but I really can't do that."

To decline again, say "I'm sorry, no." If that isn't firm enough, give him an outright "No." Use a downward inflection in your voice for a soft delivery.

Non-Monetary Solutions

When money is a barrier but he will not budge, see if he can offer anything else to sweeten the deal for you. Ask for things that will cost him little or nothing, but that have value to you.

Other People

It's often the case that you are not negotiating with the only person (or people) the deal will affect. This is (usually) a hidden barrier that can cause problems later.

Pre-empt this by asking how the deal will affect other people concerned. Find out if they are in favor of it and/or what objections they have.

Another people-related barrier is the introduction of new negotiators. This almost always means your opposition plans to take a harder line. If this happens, start from where you left off. Reiterate what you have negotiated so far, use active listening, and overcome any new barriers.

Related Chapters:

- Active Listening

BARGAINING

When active listening and your tools for overcoming barriers are not effective, you need to resort to bargaining.

Bargaining is also a good go-to for quick price negotiations, as in a street market.

The Ackerman model is a bargaining strategy with a number of psychological tools built into it. Although it's based on monetary negotiations, you can adapt it to other things too. This model assumes you're trying to get something for a cheaper price (that is, you are the "buyer"), but it also works if you're the "seller."

Step 1. Set Your Target Price

Make it an ambitious but possible number. It must not be a round number. Instead of $500, for example, use $497.98.

Step 2. Set an Extreme Anchor

Let him make the first offer.

Counter with 65% of your target price, assuming his offer wasn't better than that.

Making a super-low first offer lowers your opposition's expectations and gives you room to move. He may try the same thing with his first offer. Do not let that change yours. Stick to your plan.

You can pre-empt an objection to your low (or high) price by referring to other cases, such as the online cost or how much another company would charge.

Step 3. Raise Your Offer in Decreasing Increments

Your second and third offers should be 85% and 95% of your target price, respectively, with your final offer being 100%.

Raising your offers in this manner (65, 85, 95, 100) gives the impression you're being squeezed.

Make sure you use all your tools to overcome barriers and say no before each increase. Never increase your offer before the other person has made a counter-offer.

Step 4. Implement or Walk Away

Sometimes people need a little extra push to close the deal. Use a call to action, like "Let's do this". If that isn't enough, show your opponent what he'll lose if he doesn't move forward. People are more likely to take action if there is a loss as opposed to an equal gain.

Once you settle on terms, review them and devise an implementation strategy (if applicable). A good way to do this is with a "how" question, such as "How would you like to get this done?"

If you aren't happy with the deal, walk away from it.

Related Chapters:

- Active Listening
- Overcome Barriers

GATHERING RESOURCES

Once you have gathered information and created an escape plan, you need to get your hands on things that will help you escape and, if applicable, help you survive once you have escaped (if you're being held captive in the wilderness, for example). At the same time, you don't want to have to carry too much stuff, as it will hinder your escape.

USEFUL ITEMS

Ideally (but not likely), you will have a backpack filled with every-thing you need. Here are some things to consider, with examples in parentheses:

- Defense (gun, knife).
- Escape (breaching tools, distraction devices).
- Navigation (map, compass).
- Fire (matches, flint and steel).
- Water (water bottle, filter, purification tablets).
- Food (candy bars, jerky).
- Shelter (poncho, survival blanket).
- Signaling (mirror, whistle, flashlight, cell phone).
- First Aid (bandages, antibiotics, iodine).

Procuring most of these things will be near impossible when captured, but once you escape, you can find and/or improvise them while on the run. To learn more about how to do this, visit:

www.SFNonfictionBooks.com/Evasive-Wilderness-Survival-Techniques

Even in captivity you can gather useful items, such as nails, pieces of glass, or cordage. Don't discount anything until you have considered all possible uses for it.

If you get your hands on a pen, draw a map on the inside of your clothes.

Eat any extra food you are given in captivity to regain your strength. Once you are healthy, start stockpiling for your escape or if your captors stop feeding you.

Hobo Roll

If nothing else is available and you have the resources, make a hobo roll to carry your things. Get a piece of material approximately 90cm (35in) squared. Strong, waterproof material is preferred.

Place two small stones in opposite corners and fold the corners of the cloth over the stones.

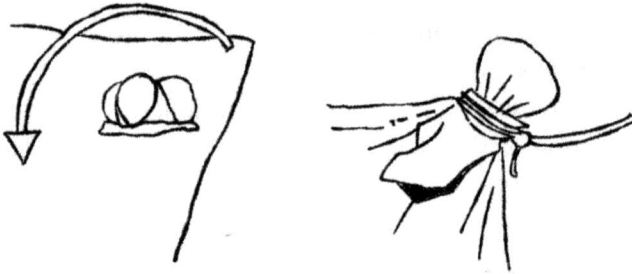

Lay the cloth on the ground and lay your things along one edge. Place the most-used items on the outside and pad hard items. Roll your things up in it tightly.

With a length of cordage, tie each end below the stones and then wrap the whole thing around your body in a comfortable position for you to carry.

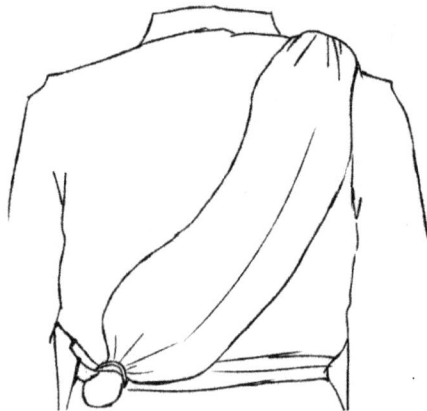

PICKING POCKETS

Knowing how to pick pockets can help you to gather resources from your captors or in the street when you're on the run. These lessons will also give you protection against pickpockets.

A successful pickpocket is a gray man. He is someone that other people overlook and don't worry about. Achieve this non-suspicious persona, and your ability to pick pockets will improve.

Choose a Mark

A mark is a person you plan to pickpocket. He is your victim.

Choose the person with the most value. On the street, this would be someone with a lot of money and/or car keys. As a captive, you might choose someone who has the keys to your cell or a weapon.

The only way to determine who has what is with observation. If you need money, go to places where money is visible, such as ATMs, racetracks, bars, and banks.

Older people are easier marks because they often need help, which allows you to get close. They are also less sensitive to your moves.

Once you have a mark, tail him until an opportunity presents itself.

Determine Location

Never attempt a lift (the act of taking something) unless you know where an item is.

Watching where your mark puts an item of value is the most reliable way to confirm its location. Another option is to look for the weight and/or shape of the object. Your mark may check it periodically, placing his hands on it to make sure it's still there.

The back pocket is the easiest to lift from. A skilled pickpocket can get to any pocket, but as an amateur, you should avoid:

- Tight pants.
- Front pockets.
- Inner jacket pockets.
- A wallet placed side-on (that is, in a position where the fold of it is not directly up or down).

Wait for or Create an Impact Distraction

Pick your mark's pocket is when he is distracted. People can only concentrate on one thing at a time. You want him focused on anything that isn't you or what you want to take.

The best type of distraction is one that has a small physical impact with him. This is because any sensation from a greater force nullifies that of a lesser force. For example, if someone taps him on the shoulder, he is less likely to feel you removing his wallet.

Impact distractions can occur naturally in crowded places, or you can create them. For example, you can spill something on him (preferably something hot) or bump into him while pretending to be drunk.

Lift the Item

The two-finger snag is an easy way to lift items of various shapes and sizes, such as a phone, keys, or a wallet, especially if they are in a rear or outer coat-pocket.

Stand behind your mark and make a narrow "V" with your index and middle fingers.

Place your fingers inside his pocket, just far enough to touch the item, but no more.

Jerk it out quickly and forcefully as the distraction occurs.

When you have the time, such as when you're waiting in line, you can nudge a phone or wallet up bit by bit. Make sure your hands are visible after every big push.

Aborting

If a suspicious mark turns around, throw the wallet on the floor and then pick it up, saying "I think you dropped this."

When caught before the lift is complete, play it off as an accidental bump.

If accused, deny everything. Run if needed.

Practice

Pickpocketing is a skill, and like any skill, it requires practice for you to get to be any good at it.

Practicing on real marks is not a good idea, but practicing on real people is essential, as they can provide feedback. If you need a cover story, tell them you are learning sleight-of-hand magic.

When a real person is not available, use a mannequin, a coat on a chair, and/or trousers filled with rags.

Related Chapters:

- Common Scams and Petty Theft

OPENING LOCKED BAGS

Opening a locked bag may uncover some useful stuff that you can use to escape.

Opening Zipped Bags

To open a locked zipped bag, move zippers all the way to one end. Use a ballpoint pen to break the zipper track (jam it between the teeth) and get what you need. Re-zip the bag, and it will work as normal.

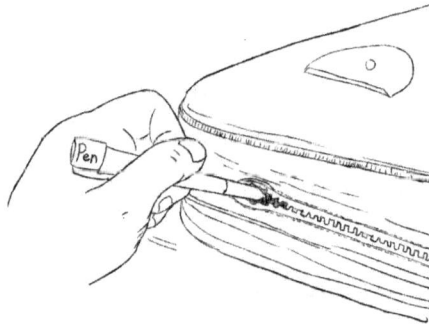

Luggage Locks

You can bypass these small, cheap, locks with a paperclip.

Bend one end of the paperclip into a small loop. Insert the loop into the lock and move it around until you find the clamp. Rotate the paperclip until the lock pops open.

ESCAPING RESTRAINTS

Until you're in a secure location, your captors will probably restrain you.

Here are a bunch of techniques to escape common types of restraints such as duct tape, zip ties, rope, and handcuffs. The technique you use depends on the material used to restrain you.

POSITION AND WRIGGLE

To position yourself for an easier escape, present your hands in front of you and make yourself larger by:

- Puffing your chest.
- Flexing your muscles.
- Pushing your forearms down so the restraints are around the larger part of your arms.
- Spreading your hands while keeping your thumbs together. This creates the illusion of closed palms, while leaving a gap at your wrists.

Once you are tied up, shrink yourself down to normal size to create gaps you can wriggle out of.

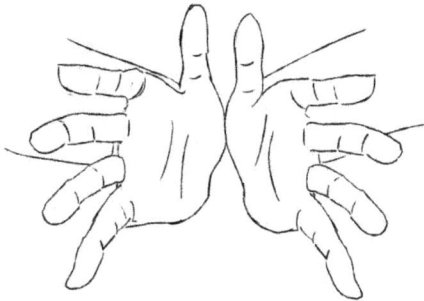

When in a chair, breathe in deeply and arch your lower back.

Straighten your arms as much as you can without being obvious, and move your feet to the outside of the chair legs.

If possible, bunch up a bit of the rope in your fist.

Once alone, do not reposition yourself until you have accessed the restraints and made an escape plan. You do not want to make things worse or get caught in the act with no plan.

If your hands are behind you, move them to one side of your body and look down. Alternatively, use any reflective surface, such as a window or mirror.

To get your hands in front of you, lower them to the back of your knees and step through them one at a time.

To wriggle out of rope, straighten your arms out in front of you and press your hands flat together. Shimmy your arms back and forth until you can get one out.

CUT

Many types of restraints are easy to defeat if you have something to cut them with, such as a razor blade, glass, or an aluminum can. Be careful not to cut yourself, especially in an artery.

When you don't have something sharp, find a 90-degree angle. A rough surface, such as the corner of a wall, a chair, or a piece of furniture, works best. Put your hands in the middle of the edge and do a sawing motion until the material is cut.

If you have some paracord, tie a foot loop on each end. Insert the paracord between the restraint material and your body. Place your feet in the loops and lie on your back. Make a bicycling motion with your feet to saw through the restraints.

FORCE

Momentum and force will tear duct tape. If you need to do so, take these actions suddenly:

To free your ankles, turn your feet outward into a V. Squat down quickly, driving your buttocks into your heels.

For your wrists, extend your hands forward at shoulder height and then drive your elbows back past your rib cage.

An alternative method to force your wrists free is to raise your arms high above your head, and then drive your arms down and out to the sides, past your hips.

To use this method for escaping from zip ties, first move the locking mechanism to where the palms of your hands meet, or as close as you can get.

When duct-taped to a chair, lean back as far as possible.

Thrust your head towards your knees as if you were assuming the crash position on an airplane.

For handcuffs, use a thick piece of metal (like a seatbelt) to pry the arms apart and break the rivet. Expect to get cut while doing this.

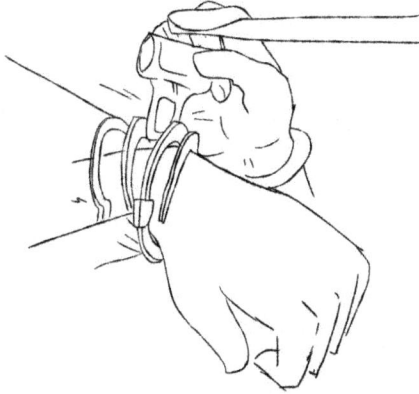

Related Chapters:

- Cut

SHIMMING

Use any thin wire (e.g., a paperclip) to shim open the locking mechanism of zip ties.

Wedge the wire between the ratchet and teeth of the zip tie, and then pull your wrists apart.

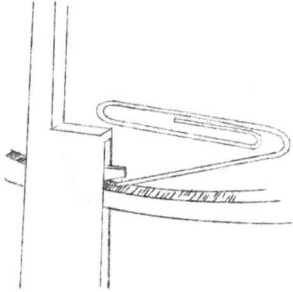

You can use the same principle on handcuffs that aren't double-locked, but the shim needs to be sturdier. A bobby pin or hair barrette works.

Drive the shim between the teeth and the ratchet. Once it is in as far as you can get it, tighten the handcuffs on you a little more to get the shim in deeper. This will release the cuffs so you can lift your hand out.

Ensure your shim isn't too thin/weak, or it will get stuck. For example, if using an aluminum can, double it up.

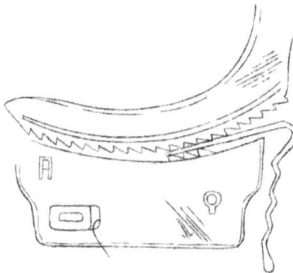

PICKING

Picking is a safer way to get out of handcuffs because you don't have to tighten them. It also works on double-locked handcuffs.

Use a bobby pin or thick paperclip to create the pick.

Straighten it out and make two 90-degree bends. If using a bobby pin, make the bends on the smooth side of it.

You can use the keyhole in the cuffs to make the bends.

Hold your hand so the teeth of the cuffs are on the bottom. Insert the bent end of your pick into the small slot part of the keyhole until it hits metal.

Pull it towards the ground and to the right in two distinct movements. This action does not require force.

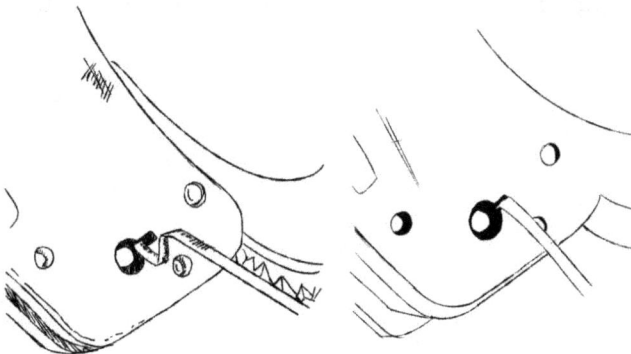

For double-locked handcuffs, release the other side first in the same manner.

SURVIVE BEING BURIED ALIVE

Getting buried alive is a grim death and escape is unlikely, but you may as well try.

If you are in a coffin, your oxygen is limited, so you need to escape as soon as possible. Try not to panic, and conserve air by taking deep breaths and holding them for as long as possible.

Use any hard object you have to tap SOS on the roof.

The other option is to try and break out. Feel for where the wooden boards combine and scratch the coffin in that place with something hard to make it easier to break. You want to make a small crack for some dirt to fall through. This will loosen the dirt above it.

Remove your shirt and close the bottom with a knot. Put your head through the neck hole so it is inside the shirt. This will protect you from suffocating.

Use your feet to push up against the coffin lid to break it.

When the dirt starts rushing in, use your hands to direct it below your feet. Aim to fill up the coffin with the dirt. Once the coffin is full, start digging up and try to stay in the bubble of air until you get to the surface.

ESCAPING ROOMS AND BUILDINGS

To break out of captivity you will need to breach doors, gates, and/or windows.

This information is also useful for getting into places in case you need to hide while evading your captors. Finally, you can use it to bolster the security in your home.

Only practice the techniques in this section on your own property. Otherwise, you may find yourself captured by the police!

RESTRAINING A GUARD

After taking out a guard, take what you can and restrain him so he can't chase you or alert others when he wakes up. These methods are also good for restraining an intruder until the police arrive.

Arm Restraints

Always tie a person's wrists behind his back. Position his hands knuckles-to-knuckles, with his palms open. If you have enough material, tie around his elbows too. Once those are secure, adapt these methods to his ankles and knees if you can.

To use paracord or something similar, tie a prusik knot on your finger. The Escaping from Heights chapter has more information on prusik knots.

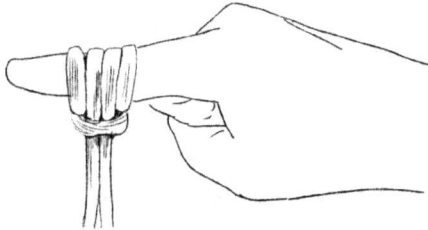

Insert the running ends into the prusik and tighten them to create loops.

Put one wrist through each loop and tighten the loops.

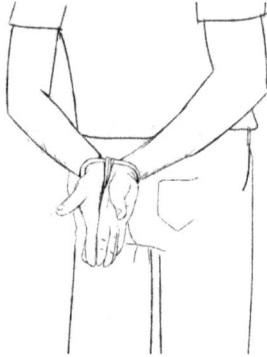

To use zip ties, chain two of them together loosely.

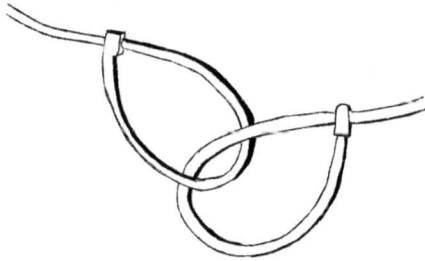

Tighten one zip tie around each wrist.

To use a belt or something similar, clasp the person's wrists together with the belt looped through the buckle. Pull it tight and then lash the remainder of it between the wrists.

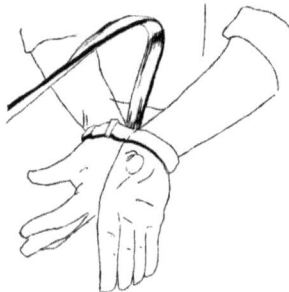

Use duct tape and rope in the same way. Secure the wrists together and then lash it.

Tying to a Chair

A chair with an open back is preferable.

Make the prisoner sit on the chair. Have him thread one arm through the back (if possible) and the other one around it. If there is no gap for him to thread his arm through, make him wrap both of them around the back. Tie his wrists together, then tie his upper arms to the chair, one on either side. Do the same with his feet, allowing only his toes to rest on the ground.

Gagging

To keep your prisoner quiet, force cloth into his mouth. Use at least two strips of tape over his mouth. Do not cover his nostrils.

Conscious Prisoners

If the guard is not unconscious, but you have a gun on him, keep your distance so he can't grab you (or your gun), and give him clear instructions. Keep calm and be prepared to use your weapon. Give him the following commands:

- "Hands up."
- "Turn around."
- "Lie down on your stomach."
- "Look away from me."
- "Hands behind your back."
- "Cross your ankles."

Alternatively:

- "Hands up."
- "Face the wall."

- "On your knees."
- "Chest and face to the wall."
- "Look away from me."
- "Hands behind your back."
- "Cross your ankles."

From this position, knock him out with your weapon by striking him as hard as you can at the base of his skull.

The only way you can restrain him while he is conscious is if you have a second person. In that case, keep your weapon aimed at him while your partner applies the restraints. If you are the one applying the restraints, control his head with your knee as you put them on.

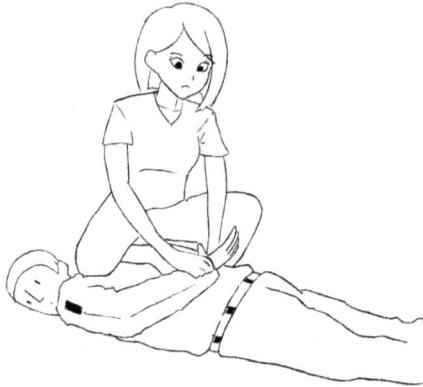

Apply an arm bar if you need to move him.

Tying to a Tree/Pole

This method does not require you to have any material to restrain him, but he does need to be conscious.

Tell your prisoner to climb the tree or pole.

Get your partner to place the prisoner's right leg around the front of the tree so his foot ends up on the left side of it. Put his left leg over his right ankle and then put his left foot back behind the tree on the same side as his body. Force him down so his body weight locks him in place. In this position, he will cramp up within 15 minutes.

To release him, you need three people: one to guard and the other two to release him. With one person on each side, lift him up by his legs and unlock them.

Related Chapters:

- Escaping From Heights

LOOK FOR THE EASIEST PATH

Before breaching any single exit point, look for the easiest way out. There may be an open window, crawl shaft, or some other unlocked point close by.

Check for other vulnerabilities as well. For example, a gate may be padlocked, but the post may not be secure, or there may be a key lock box that is easier to access than the door.

Look for the key in desks, on the door frame, under the door mat, or hidden in/under things near the door (flowerpots, rocks, etc.).

Another thing to consider is social engineering. You can follow people through doors, go through service elevators, etc. The trick here is to look the part. Act like you are supposed to be there, and you are less likely to be caught out. Having a fake badge or ID will help, as people are conditioned to see a person with a badge as a worker who is meant to be there.

When a covert escape isn't possible, wait until your captor opens the door and take him out. Hide behind the door or look weak, and when he gets close, attack.

Related Chapters:

- Elevators

DOORS AND WINDOWS

Doors and windows are obvious exit points, and will probably be locked and/or guarded.

Entering a Door

Before opening any door, listen for sound. Move the door very slowly, and never stand in front of it.

If the door is closed, approach it from the latch side. Press your back against the wall and slowly open the door a tiny bit. Ensure light or shadow doesn't give you away, and peek through the gap. If you don't see any danger, continue to open it little by little until you're confident it is safe to enter the room. Close the door carefully behind you.

If a door is already slightly open, approach it from the hinged side and peek through the crack.

Sliding Doors/Windows

Sliding doors and windows often have simple latch locks that are easy to bypass.

One method is to push against the door or window. Apply pressure while lifting and dropping it a few times. This can cause the lock to fail so you can slide it open.

If you have a thin piece of wire, slide it between the frame and the latch to unhook the latch.

A more forceful method is to slide a lever, such as a screwdriver or crowbar, between the frame and the lock and pry it open.

Jump some sliding glass doors or windows out of their tracks by prying up and outward. Once they come out, catch them before they fall.

Remove any stoppers (dowel in the frame) by creating a gap with your prying tool and using a long wire, such as a coat hanger, to maneuver them out.

A fast but noisy solution is to smash the glass. Cover the point of impact with a folded-up towel or something similar to muffle the sound. Do not use your body to break it.

Breaking Through Doors

A door is only as strong as its weakest point. Several well-placed kicks where the lock is mounted is often enough to bust it open.

If you have channel-lock pliers, use them to twist the lock until the retaining bolts break. Use a knife or something similar to turn the bolt.

You can pry a door open with a crowbar by inserting it between the lock and the door and maneuvering it back and forth.

When the hinge pins are on your side of the door, knock them out with a hammer and nail.

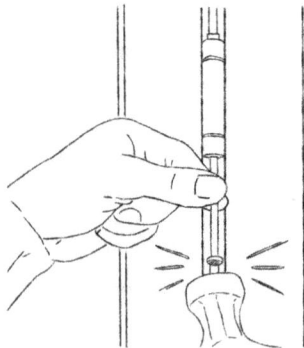

Many interior doors have a small hole or keyhole on the doorknob as an emergency unlock feature. Insert a probe, such as a paperclip, and push or turn it to release the lock.

Coat-Hanger Breach Tool

You can bend a wire coat-hanger in specific ways and then feed it through gaps to bypass locks. For example:

- Push the crash bar of an emergency exit door down.
- Lift the wooden dowel in a sliding window/door
- Nudge simple push-lever locks, such as those in car doors, down.
- Pull down the handles on the insides of auto-locking doors, such as those in hotels.

Here is an example of a DIY crash-bar breacher.

PADLOCKS

Most padlocks can be forced open by using a hammer, large stone, or brick and smashing down where the shackle meets the body on the side of the locking latch. If you can't see the locking latch, do it on both sides.

You can also shim padlocks, especially the low-quality ones.

To make an improvised padlock shim from an aluminum can, cut out two rectangles with a semicircle knob. The exact size you need depends on the size of the padlock. With practice, you will be able to guess by looking at the lock.

Fold the base up to increase strength.

Shimmy the semicircle between the bar (the shackle) and the base of the padlock.

Once both shims are in, rotate them so the handles face outwards.

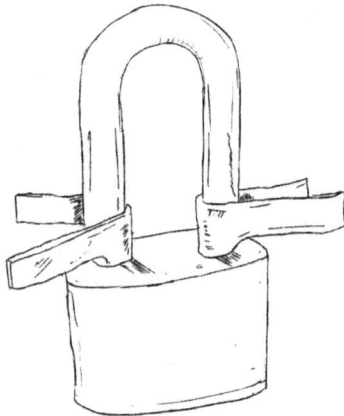

Pull up on the shackle to open the padlock.

This also works for dial-style combination locks.

Some padlocks are "anti-shim," but no lock is impenetrable. If there is a specific lock you want to get through, search it on YouTube and there may be a tutorial.

Unfortunately, in a capture scenario, you won't have access to the internet and the tools needed.

Cracking Combination Locks

This method is for combination locks without false gates, which are generally cheaper locks.

1. Apply constant pressure on the shackle away from the body of the lock.
2. Test each number to see which one gives the most resistance.
3. Once you identify the one with the most resistance, turn it until you hear a click. and feel the body move down a little. If it clicks but doesn't move, it isn't right.
4. Repeat steps 2 and 3 for each number.
5. For the last number, release the pressure on the shackle and use trial and error. Go through each number one by one until it opens.

You can adapt this same technique for chain-style combination bike locks.

Freeze a U-Bolt Lock

U-Bolt bike-locks are notoriously tough, and unlike with most padlocks, smashing them with a hammer will probably not open them.

To weaken the structure of the lock (or any metal), use a keyboard-cleaner air compressor to freeze it. Hold the can upside down and spray where the bar meets the lock until it is frozen. You may need a few cans to do it.

Smash it with the hammer until it breaks off.

SLIPPING LOCKS

When a top lock (also called a Yale lock, a night-latch lock, and other things) has not been locked by the switch at the bottom and opens inwards (like most external doors do), you may be able to slip it.

Here is what this type of lock looks like. It's characterized by the front round lock (circled).

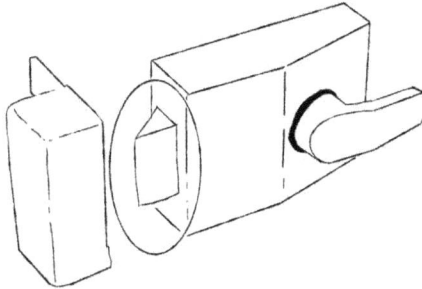

In old movies, you often see people slipping locks with their credit card. Don't do that. Your credit card will probably break. Instead, use a slim sheet of plastic a bit larger than an average adult's hand. Plastic soda or milk bottles cut into a rectangle work well.

Insert the plastic between the frame and the door, just above or below the lock. Move the plastic towards the lock until it hits the latch. Keep pressing the plastic against the latch as you gently pull the door towards you.

When the latch is pressed in, by the plastic you may hear a pop or click sound.

You can also slip a lock with a large paperclip and a shoelace. Straighten the paperclip and loop the shoelace around it so about 4/5ths of the shoelace is wrapped around the paper clip. Curve the paperclip into a rough U shape.

Feed the paperclip behind the latch and back out, so that the shoelace is wrapped around the latch but you have both ends. Pull the lace and the door at the same time to slip the lock. Double doors (e.g., French doors) are particularly easy to slip.

If there is no room to slip the lock, you can use a screwdriver or something similar to make a gap between the lock and door.

You can also "slip a lock" over time, though technically it is not slipping the lock. Stuff bits of paint (or whatever) into the strike plate every time you pass through. Eventually, they will block the latch enough to keep it unlocked.

Modern top locks come "anti-slip," but they are often installed incorrectly. In your own home, use a deadbolt instead.

PICKING LOCKS

This works for most pin-tumbler and wafer locks, which is how most key locks operate.

It is good to know how a pin-tumbler lock works. Here is a basic description:

A pin tumbler lock consists of two rows of pins held down by springs. There is also a shear line.

When the correct key is inserted into the lock, it pushes the top pins up to clear the shear line. A bottom pin will break off from the top one, "setting" the pin. Once all the pins are set, you can turn the lock.

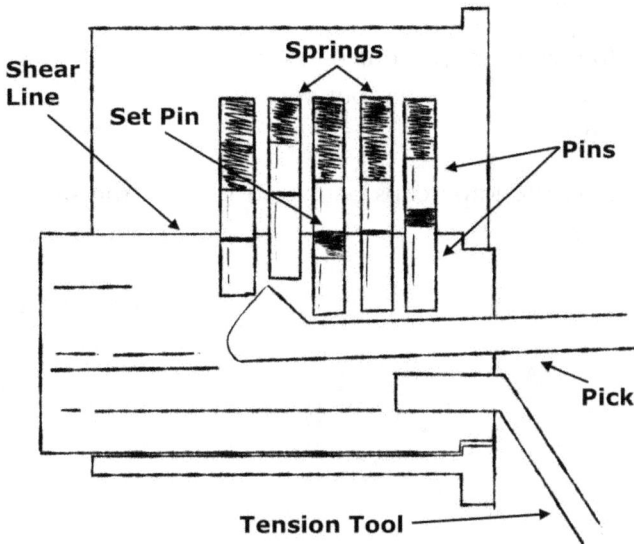

To pick a lock you need to apply slight tension on the rotation (with a tension tool) and then move each of the pins into their correct place. The tension keeps the pins in place as you move them.

Wafer locks (found in cupboard doors, file cabinets, old padlocks, and other places) operate differently, but you pick them in the same way. They are generally easier to open than pin-tumbler locks.

Creating Lockpicking Tools from Paperclips

When first starting to learn, you may wish to purchase proper lock picks, but if you get captured you probably won't have these on you. It is illegal to carry lock picks in many places, and even if you do, your captors will probably take them off you.

Paperclips are easier to hide and standard security won't confiscate them.

Bobby pins work well as tension tools, but are a bit thick for picks. However, it is possible to use them, so if a bobby pin is all you have, you may as well try.

Make the following shapes out of your paperclips. A pair of pliers will make construction easier, but you can do it without them in an escape scenario.

Avoid bending the paperclips back and forth at the same place, as they will break.

C-Rake   **Tension Tool**

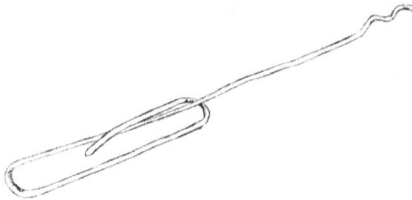
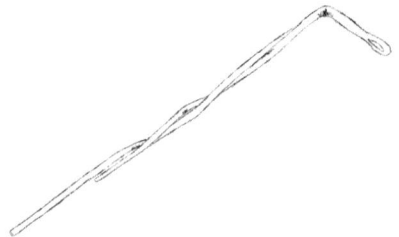

Flatten the ends by sanding them down on the floor or wall. This will allow more room for maneuvering and ensure that both tools are able to fit inside the lock simultaneously.

Raking the Lock

Raking is the fastest way to pick a lock, if it works.

Insert the tension tool into the keyhole at the point furthest away from the pins (usually at the bottom) and apply slight rotational pressure in the same direction that the lock turns.

After practice, you'll be able to feel which way the lock opens by turning it with the tension tool. You'll feel slightly less pressure when turning it in the right direction.

Most people find it easier to use the tension tool with their non-dominant hand.

The biggest mistake beginners make when learning to pick locks is using too much rotational pressure on the tension tool. You only need a small amount of tension. It's also important to keep the pressure on the tension tool consistent while you pick the lock. Don't add additional pressure until all the pins are in place and you're opening the lock.

Once you have the tension tool in place, insert the c-rake into the lock.

Lift and pull it out in a fluid motion. Move your rake in and out of the lock in this motion until the pins are "bounced" into position and the lock turns open. The rake is always in the lock. Do not pull it completely out.

Some people do this with a rapid in-out motion, and some people prefer to do it slower. It will depend on you and the lock. Either way, always lift up and out on the way out, and do it fast enough for the motion to be smooth.

If the lock doesn't open after several attempts, it's probably because of too much or not enough tension on your tension tool.

If you want to see some videos, search "raking a lock with paperclips" on YouTube.

Picking the Lock

If raking isn't working, you can try picking the lock. To pick a lock, you need to raise each pin in place using a feeler pick instead of the c-rake. Expect at least five pins.

This is the shape you need to make the paperclip. The tension tool is the same as before.

Place the tension tool in the lock, the same as you do when raking.

Insert your feeler pick with the raised bump facing the pins, which is usually away from the tension tool.

Start at the front or back of the lock, and use the bump in your pick to raise each pin in turn until you identify the stiffest one. Lift this pin until you feel it set in place. There may be a slight give or click. It's hard to explain, but with practice, you'll know.

Repeat this for the next stiffest pin, then the next, and so on for all of them.

Once all pins are in place, you'll feel the tension tool give a little, and you may hear a click. Apply more pressure on the tension tool to open the lock.

When you push a pin up too far and it gets stuck, you have over-set it. Try releasing a bit of tension or jiggling the pick. If that doesn't work, you need to start again.

If the pins keep falling, you need a little more pressure on the tension tool.

Racking and Picking Combined

You can use raking and picking together. Rake the lock to set whatever pins it has and then use the feeler pick to do the rest. Often, the pin at the back will need the extra attention.

Dummy Pins

More secure locks can have dummy pins to prevent you from picking them. The most common is a spool pin.

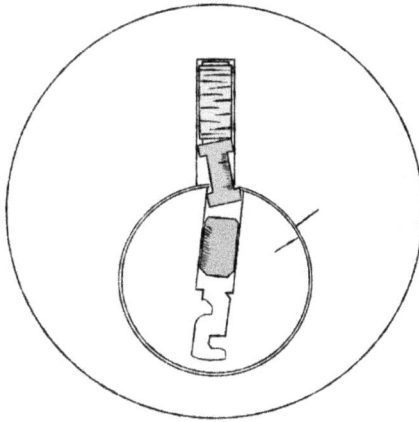

This design can trick you into thinking you have over-set a pin.

You can identify a spool pin because it has more rotational give than the regular ones.

If you think you're stuck on a spool pin, you can verify it by putting a little more upward force on it with your pick. Doing this on a spool pin will create backwards pressure on your tension tool as the bottom ridge of the pin pushes back.

Once you have identified a spool pin, bypass it by releasing a tiny amount of pressure off your tension tool and gently pushing up on the pin.

If you feel push-back pressure on the rotation as you do this, it means you're doing it right. Keep pushing until it sets as normal.

When you're setting the spool pin, other pins may drop due to the release of pressure on your tension tool. Just reset them now that the spool pin is in place.

Practice

The theory of picking locks is simple, but it takes lots of practice to get good at it.

Don't practice on the same locks all the time. Not only is it not realistic, but it will damage your lock.

Making Duplicate Keys

If you get temporary access to the key you need, you can make a duplicate.

This probably won't help you once you are captured but you never know when it could come in handy.

First you need to make an impression of the key. You can do this by:

- Taking a photo of it.
- Pressing it against your skin and then tracing the indentation.
- Pressing it into something soft that will hold an indent, such as play-dough, wax, a bar of soap, or Styrofoam.
- Making a tracing by laying the key under paper and scribbling on top. This is a last-resort method, as it is not very reliable.

Once you have the impression, photocopy it at a 1:1 ratio. The photocopy must be exactly the same size as the key, so take this into consideration if taking a photo.

Cut an outline of the key from the paper copy, and then trace it onto an aluminum can that has been cut open and laid flat. Cut the shape out of the aluminum. For greater accuracy, first cut a broad shape and then cut the detailed grooves.Use this key to push the pins in place and a tension tool to turn the lock.

Related Chapters:

- Padlocks

SENSOR LOCKS

Locks with sensors are common, and bypassing them is harder than getting around normal key locks, but not impossible if you have the right tools.

Motion-Sensor Doors

This is for doors that use a motion-sensor trigger to unlock, such as those that open from the inside but not the outside, or those where you need an ID card to get in, but not out.

Many of these auto-locking motion sensor doors use passive infrared (PIR) sensors. These can be fooled with compressed air duster cans, such as keyboard cleaners. Hold the can upside down and spray it at the sensor, and the door will open. Other things such as vape smoke or a spray of water may also work. Some doors require a temperature variance to trigger, which is why the compressed air is more reliable.

If the door is electro-magnetic, this will not work.

RFID Badge Cloning

You can clone most RFID badges or FOBs with an RFID cloner. Buy one (e.g. Proxmark) online. Conceal it in a coffee cup, sandwich bag, etc., so you can get closer to your mark without raising suspicion.

Magnetic Locks

Place a bit of black duct tape or a paperclip over where the magnets connect. This will prevent the magnetic seal from forming when it closes.

Motion Sensors

Motion sensors are not technically locks, but they may be present you're when trying to escape.

One option is to trip a sensor several times on purpose, so the owner switches it off.

To fool modern motion sensors is hard. You must study them first. Determine the area a sensor monitors and look for a path around it. Move slow and low along the walls which the sensors are placed on. Be aware of any other sensors facing the wall. Use furniture as cover to block your movement. The sensor may be calibrated for pets, so staying low is a good idea.

ESCAPING FROM HEIGHTS

When you need to escape a building from a high floor, the best thing to do is take the fire stairs. Stay close to the wall and away from the guard rail, especially if the whole building is being evacuated.

Rappelling

If you are trapped in a room, you can rappel out.

A king-size bed sheet will create a harness large enough for most adults. Other materials will also work as long as they are strong enough.

Fold the bed sheet in half to make a triangle, then roll it up from the base to the tip.

Tie the ends together with a square knot:

- Right over left and under, left over right and under.
- Pull both right ends away from both left ends to tighten it.
- Ensure there is at least 15cm (6in) in tail on both sides.

Lay the triangle on the floor and stand over it so one of the corners (not the knot) is between your legs. This is "Point 1." You face away from the rest of the triangle.

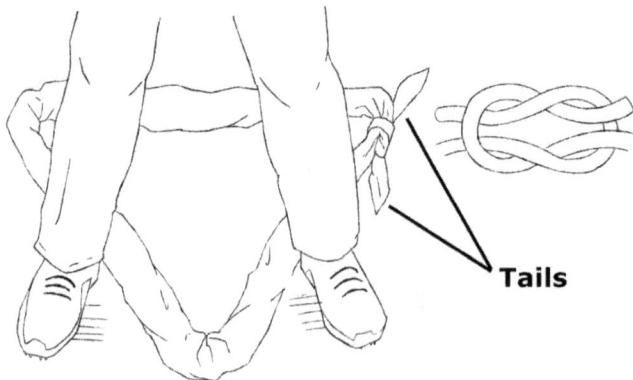

Tails

Pull the harness up so that Point 1 is at your front, between your legs, and the other two points meet it.

Next you need "rope". One king-size bed sheet is good for one story. Make the total length a little shorter than how high up you are. This way, if you fall, you will be suspended above the ground.

Tie one end to something with at least one of these qualities:

- Permanent fixture.
- Larger than the window and won't break under your weight.
- Very heavy.

Tie the sheets together using square knots (as described above), then tie the free end through all three of your harness loops.

Put some padding, such as pillows and towels, between the rope and anywhere there will be friction, such as the window sill.

Walk yourself backwards down the wall using a hand-over-hand grip on the rope. You can do this without the harness, but it won't be as safe. If you're escaping fire, wet the sheets prior to tying them and make sure the anchor isn't highly flammable.

Prusiks

Another way to climb to safety is by using prusiks. Prusiks are small loops of rope that you attach to a rope and climb up or down with. You can use them on their own or as an extra layer of safety when rappelling

They work because you can move the prusiks up, but they won't slip when you add downward pressure.

Create four closed loops. Two for feet and two for hands. If you don't know any other knots, use square knots as previously described. Others you can use, which are more reliable, are the double fisherman's knot or the figure-8 bend.

Use a prusik hitch to attach the loops to the rope:

- Put the loop across your main line, with the joining knot facing the right.
- With the knotted side, wrap your prusik loop around the main line. Do it at least twice. The more wraps you make, the more friction you will have.
- Ease the loops tight. As you do so, ensure all the lines are laid next to each other neatly. Do not let them overlap/cross each other.
- As you tighten it, do your best to position the knot close to the main line.

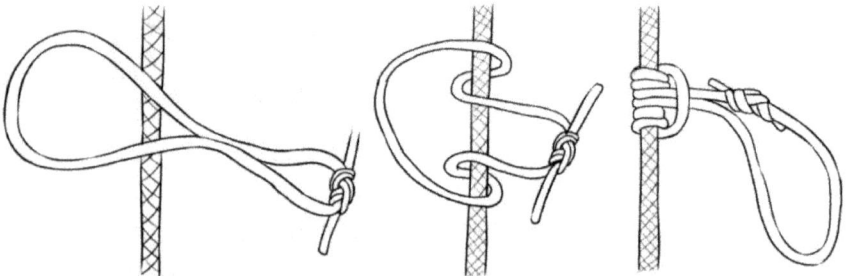

Once the prusiks are on the rope, place your feet in the bottom two loops and hold onto the top ones with your hands. Slide your hands up with the top prusik loops as high as you can, then pull yourself up. Use your legs to slide the bottom prusik loops up as high as you can. Stand up while sliding the top prusik loops up again. Repeat this process as needed.

Although it's less safe, you can do this with two prusiks (such as shoelaces) if that's all you have. Use one as a handhold and one as a foothold.

The above information was adapted from the book *Emergency Roping and Bouldering*.

www.SFNonfictionBooks.com/Emergency-Roping-Bouldering

Jumping into a Dumpster

Jumping from a window into a dumpster is a last resort, as there is a lot that can go wrong. To do this without getting a serious injury, you need:

- Something relatively soft (like cardboard) to land on in the dumpster .
- To hit the target accurately.
- To land flat on your back. Landing on your stomach can result in a broken back since your body will want to form a V on impact.

When jumping, aim for the center of the dumpster. Ensure you leap out past any obstacles, while not overshooting the dumpster. As you fall, tuck your head and bring your legs around so you land on your back.

STEALTH MOVEMENT

Stealth movement is all about getting around unnoticed. To do this you have to evade all the senses of your captors/pursuers and whatever help (e.g., dogs) they have.

OBSERVATION

Constant observation using all your senses is required when you're moving. Even when you stop, you must keep observing. Observe your enemy and/or any obstacles in your way, so you can choose how and when to move.

Searching Ground

Use this method to look for signs of your enemy, or anything else you want to look for, from a stationary position. It will help if you have something specific to look for (certain equipment, humans, dogs, vehicles, etc.).

Divide the ground into three ranges: immediate, medium, and long. Scan each section from right to left. Start with the immediate range, and work your way back systematically.

Right to left is better than left to right because we read from left to right and are more likely to overlook things if we follow that habit. Horizontal scanning is better than vertical, as that way you don't have to be continuously adjusting for distance and scale.

When you come across areas that are more likely to hide something, take a bit more time to search and look for parts of objects as well as whole ones. Things may be hidden behind something, but with bits of them still visible.

Look through visual screens, e.g., vegetation. If you want to look further, make a small head movement.

Tips for Seeing in the Dark

It takes 30 minutes for your eyes to fully adjust to the dark (night vision) and you need at least a little ambient light from a source like the moon.

Once your eyes have adjusted to the dark, you need to protect them. A flash of light can ruin your night vision in a second. When there is a bright area you want to observe, cover one eye to preserve it while you use the other one to look.

Even with your night vision, objects in the dark are harder to make out. Looking next to them will make them clearer. Changing your focal point every few seconds (up, down, to the sides) will also help.

Things may seem to move. Make sure they're staying still with the sticky finger method. Stretch a finger out in front of you and "stick" an object to it.

When you need extra light to see (if you're reading a map, for example), use red or blue light. It does minimal damage to your night vision and is harder for your enemy to spot. Don't rely solely on your vision. Sound, smell, and touch can tell you many things.

Hearing is a human's next best sense, and you can often hear things that are out of sight. Stay still, open your mouth a little, and turn your ear in the direction you want to hear.

The wind can carry smells quite far, and some smells, like food cooking or smoke, are very distinctive to humans. Turn your nose up toward the wind and smell like a dog does, taking many small sniffs. Concentrate on the inside of your nose and try to determine what the smell is.

When you can't see anything at all, it's safer to stay still until there's light, but certain circumstances may require you to move. In this case, you need to feel your way around. Move slowly, testing every movement.

Lift your feet high to give yourself the best chance of clearing any obstacles, but ensure you do not lose balance. Stretch your hands in front of you to feel for obstacles. Use the back of your hand to feel stuff, in case it's sharp or hot. This will protect the inside of your hand and the arteries in your arm.

COVER AND CONCEALMENT

Cover and concealment are different. Both are useful for stealth.

Concealment is anything between you and your enemy that hides you from sight. Vegetation is a good example of concealment. The more of it there is between you and your enemy, the harder it will be for him to see you.

Cover will hide you from sight too, but will also stop bullets. Many solid objects do not qualify as cover. Bullets will go straight through wooden fences, car doors, windows, etc.

Solid concrete, thick metal, depressions in the earth, and large trees have a much better chance of providing you cover. The more powerful the gun (or blast), the thicker the cover needs to be.

If your enemy is trying to shoot you, seek cover. If he just wants to find you, concealment is enough.

When covering ground, move from cover (or concealment) to cover, stopping at each one to observe. Make sure you know your next place of cover or concealment before leaving your current one.

CAMOUFLAGE

Having a good understanding of the principles of camouflage will help you in all areas of stealth movement. Most of these things are intertwined. Use them together for the best results.

Shape

The human shape (or anything) is distinctive, but there are ways to distort it. For example, you can attach local vegetation to yourself or adjust your posture.

Size

The bigger things are, the easier they are to spot, and the harder they are to hide. You can make yourself smaller by getting lower to the ground and/or standing sideways for a slimmer profile.

Silhouette

When an object contrasts against a plain background, the shape of its outline is its silhouette. This is most prominent when there is a dark object on a light background, or vice versa. Examples of plain backgrounds in nature are the sky and the sea.

Even a slight shade difference is enough for a keen observer to spot a silhouette. For example, wearing black clothing creates more of a contrast at night than dark blue clothing does.

To minimize your silhouette, keep to low ground and/or lower your physical profile.

Color and Texture

Every environment has certain colors and textures, and if you don't mimic those, you'll stand out.

Contrasting colors, like light-colored hair in the forest or black clothing in the snow, stand out more.

Textures may be rocky, leafy, smooth, etc.

Distort your color and texture and that of your equipment with things like mud, vegetation, charcoal, or cloth. Consider depth of features. Use lighter colors on shaded areas (around the eyes and under the chin) and darker colors on features that stick out more (forehead, nose, cheekbones, chin, and ears).

When using vegetation to blend in, ensure its color and texture continue to match the terrain as you move, since the vegetation will change and the leaves will wilt.

When you need to hide yourself quickly, lay down flat and cover yourself with foliage.

Shine and Reflection

Shine is anything that reflects light, including oily skin. An enemy can spot shine from great distances if the angle of light is correct.

Cover glass, metal, and anything else that shines (zips, buckles, jewelry, watch faces, etc.), no matter how small it is. If you need to wear glasses, line the outsides of the lenses with a thin layer of dust to reduce the reflection of light.

Reflection isn't a big deal at a distance, but can give you away if you're careless. Avoid mirrors, glass, and anything that gives a reflection. Stay outside the field of reflection—by crouching under mirrors, for example.

Light and Shadow

Avoid moving in and using light to see as much as you can, especially at night time.

Moving under or near light makes you more visible and casts your shadow. This can give you away even when the rest of you is hidden.

Always be aware of where your shadow falls, and keep in mind that the direction of the shadow will shift with the movement of the sun or other changes in light.

Knock out lights (trip fuses or break globes) if doing so won't give away your position.

The outer edges of the shadows are lighter and the deeper parts are darker. Keep in darker parts of the shadow when possible.

Your silhouette may still be seen against lighter shadows, so keep low and still until you need to move.

If you must use a flashlight, cover the head of it with your hand. If possible, use a colored lens filter.

Noise

When you're close to your enemy, you must be careful the noise you ma. The slower you move, the quieter you can be.

Ensure there's nothing on you that will rattle, jingle, vibrate, ring, or chime. If possible, jump up and down and listen for any noise you make, and secure anything you need to.

When you have the choice, keep to quieter surfaces, such as bare earth, flat concrete, wet leaves, and large rocks.

Time your movement to coincide with ambient sounds (passing traffic, barking dogs, rain, or gusts of wind) conceal yourself.

If you hear a noise that might be your enemy, freeze and observe. Get to the ground or behind cover if you can do it without getting spotted.

Use noise and movement to distract an opponent. For example, throw something in the opposite direction from where you want to go, so your enemy's attention will focus on it.

Put down small objects by touching your hand to the surface first, then lowering the object down.

Scent

Humans have certain smells (soap, food, body odor). Do the following to lessen your scent:

- Wash yourself and your clothes without using soap.
- Avoiding-strong smelling foods like those with garlic and spices.
- Don't use anything that smells unnatural, such as cologne, tobacco, or gum.
- Rub your clothes in aromatic plants (pine needles, for example) taken from your surroundings.

Pay attention if you smell the signs of humans, such as fire, gasoline, or cooking.

Keep downwind of your enemy when possible, especially if they're using dogs.

MODES OF MOVEMENT

When evading your enemy, you need to compromise between stealth and speed. What you choose depends on your circumstance, but in general, the closer you are to your enemy, the stealthier you need to be.

For maximum stealth, move low and slow. The lower you are, the "smaller" you are, and the harder you are to see.

The slower you are, the less likely you are to attract the eye and the less noise you make.

When the enemy is close, go as low and slow as you can. If he looks in your direction, freeze. You can move faster as you get further away.

There are four basic ways you can move when you're on foot.

Walk

Walking is a good compromise between speed and stealth. You can control your speed depending on your needs, and easily shift from walking into other positions, like breaking into a run or crouching down.

The basic principles of stealth-walking apply to all types of movement.

To walk as quietly as possible, place all your weight on one foot and lift your other foot high enough to clear any obstacles, but not so high that you lose your balance. Small steps are easier to control.

Test the ground by carefully pressing down on it with the outside edge of the ball of your lead foot. If the step is going to make noise —if you're stepping on a twig, for example—test a different area. On loose ground, such as that covered with leaves, you can place your feet under the foliage.

When you find a quiet spot and are ready to continue, roll to the inside ball of your foot and then to your heel, and finally to your toes. Shift your weight to your lead foot, ensure you're balanced, and repeat the process with your rear leg.

On hard ground that is noisy, muscle control becomes paramount. The slower you go, the more control you have over your muscles and the quieter you can be. You want to be able to stop at any stage of the movement and hold your position for as long as you need to.

Keep your arms and hands close to your body, ensuring they don't hit anything.

As you walk in this manner, use relaxed, normal breathing. It encourages naturalness of movement and helps to prevent gasping if you misstep or lose balance.

Wrap cloth around your feet to muffle sounds if possible.

Stomach Crawl

This is the stealthiest way to move because you have the lowest profile.

Do not slide on your stomach. That leaves too much of a trail and makes noise. Instead, use your hands and toes to do a pushup that moves your body forward. Lower yourself to the ground, move your hands up to the pushup position again, and repeat the movement.

Crawl

When crawling on your hands and knees, test the ground with your hands before applying your weight. Put your knees in the exact same place your hands went.

Run

Running while crouching is a good way to cover short distances while no one is watching. Use this technique to get past a guard whose back is momentarily turned, for example.

Going into a full run is not at all stealthy, but it is the fastest way to create distance, which is important for evasion. As soon as you're confident you are out of sight, or if you've definitely been spotted, break into a full run.

EVADING GUARD DOGS

When you first escape, you may need to get past guard dogs.

You need to take all the same precautions as you would if hiding from humans, but you also need to worry about the dogs' increased senses of smell and hearing.

Use barriers like undergrowth to disguise your scent, and keep downwind.

Approaching from an area you know other humans operate in may fool a dog as well, since it will be used to people arriving from that direction.

Disabling Dogs

There are several options for disabling a dog.

If a dog is poorly trained, giving it food can work. Put sleeping tablets (or some other drug to knock it out) in the food if you can.

Carrying an improvised dog deterrent is good for a backup in case it charges you. Some options include:

- 50/50 water with ammonia. Cleaning products are often ammonia-based.
- A compressed air can (keyboard cleaner) held upside down. It must be held upside down to get the freezing effect.
- Bee/wasp killer. This will do permanent damage.
- Bear spray.

A final option is to kill it. Fighting a dog is not easy. Expect to get injured.

- Pad at least one arm with cardboard or other material.
- As it runs at you, offer it your padded arm.

- Once it has your arm, stab it in the abdomen, either from the rear or the front.
- If you don't have a knife, smash its skull repeatedly with a something hard, like a brick.

Trying to kill a dog when you are unarmed is difficult, but not impossible.

- Once it has your arm, force the arm as far into its mouth as you can.
- Keep exerting forward pressure until you have the dog pinned on its back.
- Choke it to death by placing the bony part of your other forearm against its throat and leaning on it as hard as you can.
- Make sure it's dead. If it's unconscious and wakes up, it may attack you again.

If choking it isn't possible, or you just need to fight off a dog but not kill it, attack its weak spots. If you hurt it enough, it will probably back off.

- Kick it in the ribs.
- Yank its front legs apart to break its knees.
- Dig your fingers into its eyes.
- Kick it in the groin.
- Give it a hard blow to the nose.

OVERCOME OBSTACLES

Obstacles are any things that will slow you down as you move, and/or places where you are more likely to be seen.

Avoid obstacles whenever possible, especially ones that are inherently dangerous in themselves. The only exception is night movement. It's better to move at night, except when the terrain doesn't allow it.

Observe an obstacle from a distance before crossing it. Look for best way to cross and best time to move.

When it comes to stealth, there is an order of preference of how to cross obstacles. The one you choose depends on the difficulty of doing it and the time factor.

- **Around**. If won't add risky exposure (e.g., light, time).
- **Under**. Dig, or lift the bottom of it.
- **Through**. Find a weak point and cut a hole if needed.
- **Over**. Cross quickly and keep your profile as low as possible. To prevent injury, land on two feet and roll if necessary.

Night

When moving at night, you need to compromise between the easiest and safest routes.

Avoid using light, especially white light. Memorize your route to minimize the need to refer to your map.

A half-moon provides a good amount of light for stealth movement. It lets you see where you're going while keeping you hidden.

Stairs

Move along the edges of stairs closest to the wall. The middle will make more noise.

Around Corners

Lie flat and look around the corner. Do not expose yourself any more than necessary.

Windows and Mirrors

Stay close to the side of the building and pass below the window/mirror level.

Wire Fences/Obstacles

Ensure fences are not electrified or fitted with other security devices. Look for:

- Warning signs.
- Bare wires going into insulators.
- Small, dead animals.

To go under a wire, slide headfirst on your back by pushing forward with your heels. Place a length of wood (or something similar) lengthwise on your body so the wire slides along it. Feel ahead with your free hand to find the next strand of wire, if there is one.

When going it under isn't practical, try going through it. Cut the lower strands so there are fewer sign of tampering. To do this quietly, hold the wire near its support and cut between your hand and the support. This technique also prevents the ends from flying away.

For even less noise, partially cut the wire and finish it off by bending it back and forth. If needed, stake the wire back to allow room to crawl through

If there's a low wire obstacle, step over it carefully. To climb over taller ones, find holds near the support posts.

In the case of barbed wire, you need to take extra care you don't get snagged. Before climbing over, cover the wire with any flat, heavy material, such as:

- Carpet.
- Thick blanket.
- Several layers of cardboard.

Razor wire is very dangerous. If you absolutely have no other choice, use a curved stick to pull wire down flat and cover it with heavy material before climbing over.

Solid Wall

If you can't go around, under, or through it, find a low spot to climb over.

Test the integrity of the wall by grasping it and lightly pulling it straight down. Gradually increase strength until you're lifting your body off the ground.

Check if other side is clear (if possible), and if it is, roll over the wall as quickly as possible.

To learn how to run up high walls and overcome other obstacles, check out *Essential Parkour Training*:

www.SFNonfictionBooks.com/Essential-Parkour-Training

Open Areas

Open areas are those that have little to no cover, like grass fields. Only cross them if there is no other practical way around.

To cross open areas, choose the lowest ground possible (furrows, for example) and lower your profile as much as is practical. Consider speed vs the need for concealment.

In grass, try to time your movement to when the wind is blowing, and change direction slightly from time to time as you cross. This helps to cover up the path of your movement.

Roads, Trails, and Railroad Tracks

Never move along roads in a covert situation. To cross them, use narrow points with low traffic and concealment to minimize your exposure (bushes, shadows, a bend in the road, low ground, etc.).

Use a low run to cross them.

Be careful of areas with no traffic, as they may be booby-trapped.

Caution: If there are three rails on the railway tracks, one may be electrified.

In Public but Hostile Territory

Avoid contact with the locals, especially children and dogs. Go around populous areas if possible.

Do your best to blend in before entering. Wear local clothing, cover your skin, get clean, etc.

Unless you are fluent in the local language, do not talk. Instead, look down and keep walking past anyone who tries to engage you.

Bridges

Avoid crossing bridges. It is better to swim across. You can hide underwater and use a reed or straw to breathe.

When the body of water is too dangerous, wait for an opportune time and cross the bridge as quickly as possible.

If you are caught on the bridge and death is imminent, jump into the water. This is very dangerous, especially if you do not know the depth or the water.

When jumping, try to land in the channel where boats go under the bridge. This area is generally in the center, away from the shoreline.

Stay away from any area with pylons that are supporting the bridge. Debris can collect in these areas, and you may hit it when you enter the water.

Jump in feet-first, keeping your body completely vertical. Squeeze your feet together, clench your backside, and protect your crotch with your hands.

After you enter the water, spread your arms and legs wide, and move them back and forth to slow your descent.

Related Chapters:

- Observation

IMPROVISED EXPLOSIVES

An improvised explosive is a homemade bomb. The improvised explosives in this book use minimal equipment to give you the best chance of making them in captivity or at home.

Some of these are nothing more than simple science experiments. They're good for creating distractions.

Others are meant to hurt your enemy. I don't suggest practicing any of these, but they're good to know.

When dealing with explosives, safety is paramount. Always wear protective clothing and ensure no one—except your enemy—is in the danger zone when you're setting them off.

MATCH-HEAD FUSES

Some of the improvised explosives require fuses. An easy way to construct those is with toilet paper and matches.

Use the toilet paper to make string. Tear strips as thin as you can. Fold each strip in half lengthwise and twist it.

Next, put on plastic gloves and scrape the heads off the matches. Crush the heads so there are no large clumps. Sprinkle a tiny amount of water onto the match heads and mix it in as you do it. You want a thick paste—the smoother the better.

Coat your strings in the match-head paste and let them dry. Store the dry fuses in a paper bag, away from heat and fire.

Replacement Materials

Any string or paper can replace the toilet paper, but it may not work as well. Make sure any string you use is clean and try to get a thickness similar to that of the toilet paper string.

You can replace match heads with gunpowder. Get it from bullets.

To make an improvised gunpowder, mix the following together:

- 1 part potassium nitrate (found in fertilizer).
- 1 part granulated sugar.
- 2 parts hot water.

DISTRACTION BOMBS

Distraction bombs are easy and relatively safe to make. Set them off and when the guard(s) investigate the noise, make your move.

Although it's not technically an explosive, a fire also makes a good distraction.

Flint Flash

This "bomb" will create some small but bright sparks. It's easily missed, but if it's within a guard's field of vision, he'll probably go in for a closer inspection.

To make it, you need a disposable lighter and another source of fire.

Remove the metal flame-guard from the disposable lighter. Carefully take off the striker wheel and extract the flint and flint spring.

Twist one end of the spring around the flint. Put the flint in a flame, holding it by the spring. When it's red hot, throw it at a hard surface. The sparks are created on contact.

Lighter Bomb

Using the same disposable lighter you got the flint from, you can construct a noise-maker.

With the metal flame guard removed, move the flame adjustment mechanism until gas continuously leaks out. To do this, move it all the way to the "+." Pull it up and over, back to the "-." Repeat this action to unscrew the gas valve.

Once it's leaking, hang it upside down and light the gas.

As soon as the flame burns though, the rest of the gas will ignite, making a small explosion. This usually happens in under a minute.

Make sure you're not too close when it goes off.

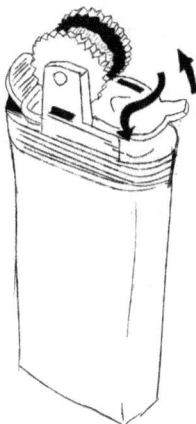

Simple Chemical Noise-maker

This noise-maker uses a simple chemical reaction to release gas inside a closed container. When the pressurized gas is released, it creates a decent bang.

You will need:

- A small plastic bottle with a lid (a water or soda bottle will work well).
- A small square of paper (such as the soda bottle label).
- 1/4 cup vinegar.
- 2 tablespoons baking soda.

The ingredient measurements do not have to be precise. Close enough is good enough, but the bigger the bottle, the more ingredients you need.

Wrap the baking soda in the paper, so it's sealed inside.

Pour the vinegar in the bottle and then drop the baking soda package in. Immediately seal and shake it. Wait for the bottle to expand a little and throw it at something hard.

To make improvised tear gas, pour chili powder and/or red pepper into the bottle before adding the other ingredients.

Other chemical reactions you can try are:

- Water + Alka Seltzer (brand of effervescent antacid).
- Coke + Mentos (brand of candy).

Works Bomb

This is a more powerful chemical noise-maker based on the same principles as the previous one (that is, a plastic bottle filled with an acid and a reacting base). It creates hydrogen gas, which is highly flammable.

The acid ingredient is hydrochloric acid. You can find this in several household agents, such as:

- Toilet-bowl or drain cleaner
- Swimming pool maintenance chemicals (muriatic acid).
- Masonry cleaner (tile cleaner).

If you have a choice, choose the one with the highest percentage of hydrochloric acid, at least 20%. A common one to use is Works brand toilet-bowl cleaner; hence the name "Works Bomb."

Be careful not to spill the hydrochloric acid on yourself. Use gloves and safety glasses.

For the reacting base, use aluminum foil.

Loosely scrunch up several small aluminum foil balls and put them in the bottle. Cover the balls with the hydrochloric acid and screw the cap on. Roll it to where you want the explosion to occur. Alternatively, gives it a couple of shakes, set it down and run away.

This may take some time to go off, but when it does, it will definitely attract attention. To see it in action, search YouTube for "Works Bomb."

VISION REDUCERS

The improvised explosives in this section are meant to impair your captors' ability to see. None of these are explosive, but they get the job done.

Flour Bomb

This works with any type of flour or other fine powder, such as ashes.

Wrap a generous amount of the flour inside a wet paper towel. Use a rubber band to hold it together. Any paper will work if paper towel is not available. A plastic wrapping can also work if the plastic isn't too thick. Pack it tightly.

Throw it at a hard surface (or at someone) to create a big puff of flour on impact.

Cooked Smoke Bomb

For this smoke bomb, you need:

- Sugar (sucrose or table sugar).
- Potassium nitrate/saltpeter (fertilizer or gunpowder).
- A skillet
- Aluminum foil shaped into a mold (any shape you want).
- A fuse (optional).

Put three parts of potassium nitrate with two parts of sugar into the skillet. The measurement doesn't have to be precise, but you need more potassium nitrate than sugar. The more sugar the slower the burn.

Place the skillet over low heat and use long strokes to stir the mixture until it's liquid. Pour the mixture into your aluminum foil mold and insert a fuse if want.

Once it cools, peel the foil off. When you want to use it, light the fuse. If there is no fuse, you can light it directly.

Uncooked Smoke Bomb

To make this smoke bomb, you need:

- 2 parts icing sugar (powdered sugar).
- 3 parts potassium nitrate/saltpeter (fertilizer or gunpowder).

Sift the icing sugar and potassium nitrate together. Light the powder to produce smoke.

FIREBOMBS

These improvised explosives are meant to be destructive, and can do serious damage.

Molotov Cocktail

This classic firebomb is a glass bottle filled with anything flammable, like liquor or gasoline.

Any cloth soaked in the flammable liquid makes a good fuse. Plug it tightly into the top of the bottle. Light it up and throw it at whatever you want to have catch on fire.

Improvised Incendiaries

Here are three ways to make a highly flammable sticky liquid— like a poor man's napalm.

Using it in a Molotov cocktail is a good way to deploy it. Use a funnel to put it in the bottle.

Mix any of the following combinations in an old container. Be careful when handling it, so you don't get any on you.

- 5 cups of gasoline + 1 cup of oil + half a bar of shaved soap.
- Styrofoam + gasoline. Use however much Styrofoam is needed until the gas can't dissolve any more.
- 2 parts flour + 1 part gasoline.

THANKS FOR READING

Dear reader,

Thank you for reading *Evading and Escaping Capture*.

If you enjoyed this book, please leave a review where you bought it. It helps more than most people think.

Don't forget your FREE book chapters!

You will also be among the first to know of FREE review copies, discount offers, bonus content, and more.

Go to:

https://offers.SFNonfictionBooks.com/Free-Chapters

Thanks again for your support.

REFERENCES

12PillarsOfSurvival.com. *Survival Stash.* 12PillarsOfSurvival.com.

Alton, J. (2016). *The Survival Medicine Handbook.* Doom and Bloom.

Auerbach, P. Constance, B Freer, L. (2018). *Field Guide to Wilderness Medicine.* Elsevier.

Carnegie, D. (2010). *How To Win Friends and Influence People.* Simon & Schuster.

Chesbro, M. (2002). Wilderness Evasion. Paladin Press.

Department of Defense. (2011). *U.S. Army Survival Manual: FM 21-76.* CreateSpace Independent Publishing Platform.

DOD United States Department of Defense. (2011). *Survival, Evasion, and Recovery.* Pentagon Publishing.

Emerson, C. (2016). *100 Deadly Skills: Survival Edition.* Atria Books.

Emerson, C. (2015). *100 Deadly Skills.* Atria Books.

Erickson, R. Erickson, R (2001). *Getaway: Driving Techniques for Escape and Evasion.* Breakout Productions.

Fiedler, C. (2009). *The Complete Idiot's Guide to Natural Remedies.* Alpha.

Goodwin, L. (2014). *Prepping A to Z: Book A.*

Goodwin, L. (2014). *Prepping A to Z The Book Series Book B.*

Goodwin, L. (2014). *Prepping A to Z The Book Series Book C.*

Goodwin, L. (2014). *Prepping A to Z The Book Series Book D.*

Goodwin, L. (2014). *Prepping A to Z The Book Series Book E..*

Goodwin, L. (2014). *Prepping A to Z The Book Series Book F.*

Hanson, J. *Don't Hide Valuables Here.* www.spyescapeandevasion.com.

Hanson, J. (2015). *Spy Secrets That Can Save Your Life.* TarcherPerigee.

Hanson, J. (2018). *Survive Like a Spy.* TarcherPerigee.

Hawke, M. Hawke, R. (2018). *Family Survival Guide.* Skyhorse.

Lieberman, D. (2018). *Never Be Lied to Again.* St. Martin's Press.

Luther, D. *The Prepper's Workbook.*

Miller, T. (2012). *Beyond Collapse.* CreateSpace Independent Publishing Platform.

Morris, B. (2019). *The Green Beret Survival Guide.* Skyhorse.

Nobody, J. (2011). *Holding Your Ground.* Elsevier.

Nobody, J. (2018). *The Prepper's Guide to Caches.* Prepper Press.

Robinson, C. (2012). *The Construction of Secret Hiding Places.* Desert Publications.

Terrill, B. Dierkers, G. (2005). *The Unofficial MacGyver How-To Handbook.* American International Press.

Voss, C. Raz, T. (2016). *Never Split the Difference.* Harper Business.

WA Police, SA. (2019). *Aids to Survival.*

Wiseman, J. (2015). *SAS Survival Guide.* William Collins.

United States Marine Corps. (2013). *United States Marine Corps Individual's Guide for Understanding and Surviving Terrorism.* United States Marine Corps.

US Marine Corps. *Kill or Get Killed.*

Yeager, W. (1990). *Techniques of the Professional Pickpocket.* Breakout Productions.

AUTHOR RECOMMENDATIONS

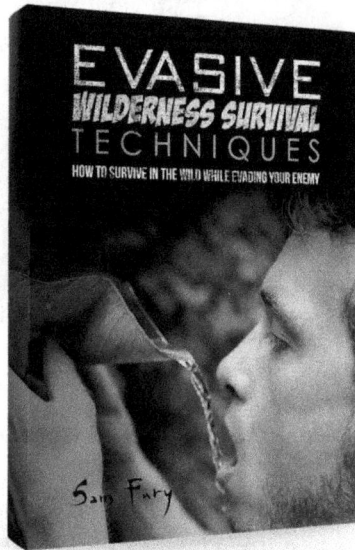

Teach Yourself Evasive Wilderness Survival

Discover all the evasive survival skills you need, because if you can survive under these circumstances, you can survive anything.

Get it now.

www.SFNonfictionBooks.com/Evasive-Wilderness-Survival-Techniques

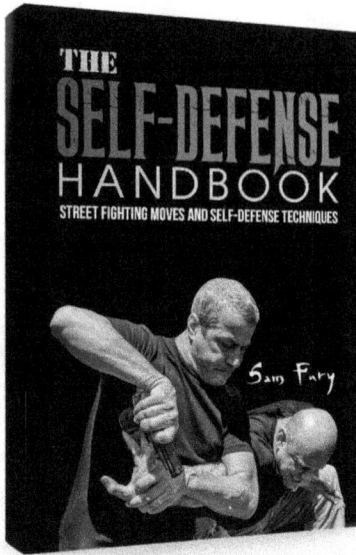

Teach Yourself Self-Defense

This is the only self-defense training manual you need, because these are the best street fighting moves around.

Get it now.

www.SFNonfictionBooks.com/Self-Defense-Handbook

ABOUT SAM FURY

Sam Fury has had a passion for survival, evasion, resistance, and escape (SERE) training since he was a young boy growing up in Australia.

This led him to years of training and career experience in related subjects, including martial arts, military training, survival skills, outdoor sports, and sustainable living.

These days, Sam spends his time refining existing skills, gaining new skills, and sharing what he learns via the Survival Fitness Plan website.

www.SurvivalFitnessPlan.com

amazon.com/author/samfury

goodreads.com/SamFury

facebook.com/AuthorSamFury

instagram.com/AuthorSamFury

youtube.com/SurvivalFitnessPlan

www.ingramcontent.com/pod-product-compliance
Lightning Source LLC
Chambersburg PA
CBHW051715020426

42333CB00014B/999